ARGOPREP

TEACHER RECOMMENDED

KINDERGARTEN COMMON CORE MATH

DAILY PRACTICE BOOK

ARGOPREP.COM

FREE ONLINE SYSTEM WITH VIDEO EXPLANATIONS

ArgoPrep is one of the leading providers of supplemental educational products and services. We offer affordable and effective test prep solutions to educators, parents and students. Learning should be fun and easy! For that reason, most of our workbooks come with detailed video answer explanations taught by one of our fabulous instructors.

Our goal is to make your life easier, so let us know how we can help you by e-mailing us at: info@argoprep.com.

ISBN: 978-1951048631
Published by Argo Brothers, Inc.

ArgoPrep has won **over 10+ educational awards** for their workbooks and online learning platform. Here are a few highlighted awards!

OTHER BOOKS BY ARGOPREP

Here are some other test prep workbooks by ArgoPrep you may be interested in. All of our workbooks come equipped with detailed video explanations to make your learning experience a breeze! Visit us at ***www.argoprep.com***

COMMON CORE MATH SERIES

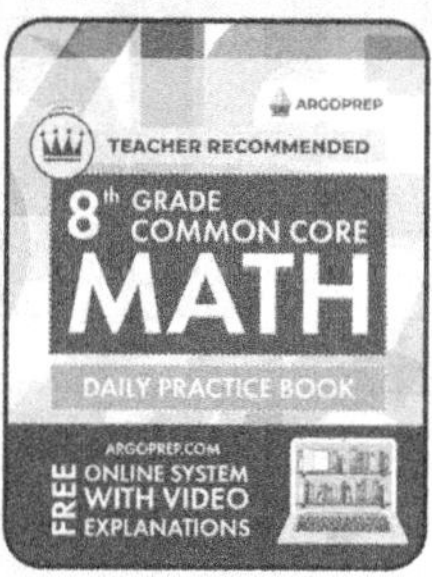

COMMON CORE ELA SERIES

INTRODUCING MATH!

Introducing Math! by ArgoPrep is an award-winning series created by certified teachers to provide students with high-quality practice problems. Our workbooks include topic overviews with instruction, practice questions, answer explanations along with digital access to video explanations. Practice in confidence - with ArgoPrep!

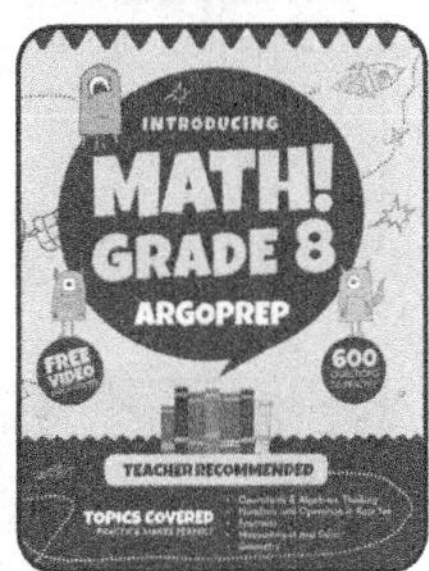

SCIENCE SERIES

Science Daily Practice Workbook by ArgoPrep is an award-winning series created by certified science teachers to help build mastery of foundational science skills. Our workbooks explore science topics in depth with ArgoPrep's 5 E'S to build science mastery.

KIDS SUMMER ACADEMY SERIES

ArgoPrep's Kids Summer Academy series helps prevent summer learning loss and gets students ready for their new school year by reinforcing core foundations in math, english and science. Our workbooks also introduce new concepts so students can get a head start and be on top of their game for the new school year!

CAPTAIN BRAVERY

WATER FIRE

MYSTICAL NINJA

GREEN POISON

FIRESTORM WARRIOR

RAPID NINJA

CAPTAIN ARGO

THUNDER WARRIOR

ADRASTOS THE SUPER WARRIOR

DANCE HERO

GREEN DRAGON WARRIOR

TABLE OF
CONTENTS

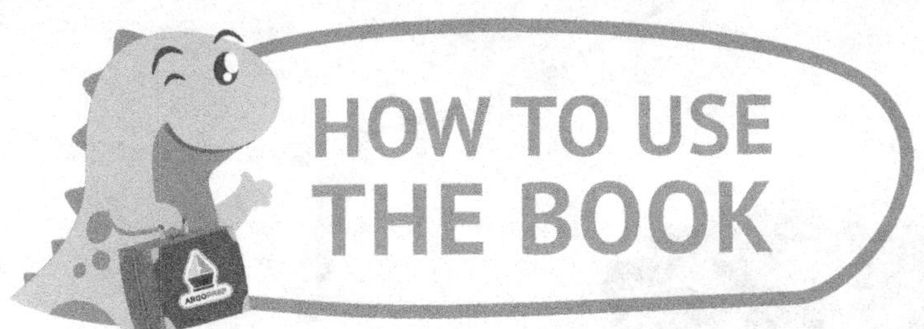

This workbook is designed to give lots of practice with the math Common Core State Standards (CCSS). By practicing and mastering this entire workbook, your child will become very familiar and comfortable with the state math exam. If you are a teacher using this workbook for your student's, you will notice each question is labeled with the specific standard so you can easily assign your students problems in the workbook. This workbook takes the CCSS and divides them up among 20 weeks. By working on these problems on a daily basis, students will be able to (1) find any deficiencies in their understanding and/or practice of math and (2) have small successes each day that will build proficiency and confidence in their abilities.

You can find detailed video explanations of each problem in the book by visiting:
www.argoprep.com/books/

We strongly recommend watching the videos as they will reinforce the fundamental concepts.

HOW TO WATCH VIDEO EXPLANATIONS

IT IS ABSOLUTELY FREE

Go to **argoprep.com/ccmK**
OR scan the QR Code:

Week 1 is all about counting to 100 by ones and by tens. Let's get started!

You can find detailed video explanations of each problem in the book by visiting: ArgoPrep.com/ccmk

WEEK 1 : DAY 1

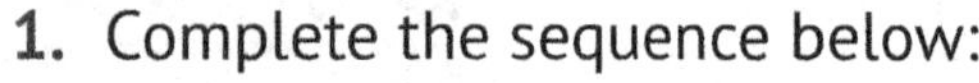

1. Complete the sequence below:

2 3 4 ☐

K.CC.A.1

2. Complete the sequence below:

15 16 17 ☐

K.CC.A.1

3. Complete the sequence below:

28 29 30 ☐

K.CC.A.1

4. Complete the sequence below:

33 34 35 ☐

K.CC.A

5. Complete the sequence below:

41 42 43 ☐

K.CC.A

6. Complete the sequence below:

56 57 58 ☐

K.CC.A

Welcome! In this workbook each page typically contains six questions. If you get stuck, you can always watch our video explanations on our website.

WEEK 1 : DAY 2

Complete the sequence below:

62 63 64 ☐

K.CC.A.1

Complete the sequence below:

79 80 81 ☐

K.CC.A.1

Complete the sequence below:

80 81 82 ☐

K.CC.A.1

4. Complete the sequence below:

94 95 96 ☐

K.CC.A.1

5. Complete the sequence below:

45 46 47 ☐

K.CC.A.1

6. Complete the sequence below:

23 24 25 ☐

K.CC.A.1

How well do you know your numbers? Count out loud from 0 to 20!

WEEK 1 : DAY 3

1. Complete the sequence below:

39 40 41 ☐

K.CC.A.1

2. Complete the sequence below:

87 88 89 ☐

K.CC.A.1

3. Complete the sequence below:

50 51 52 ☐

K.CC.A.1

4. Complete the sequence below:

4 5 6 ☐

K.CC.A

5. Complete the sequence below:

11 12 13 ☐

K.CC.A

6. Complete the sequence below:

66 67 68 ☐

K.CC.A

When counting larger numbers, it may be helpful to start by reviewing and counting from 1-10.

WEEK 1 : DAY 4

Complete the sequence below:

10 20 30 ☐

K.CC.A.1

4. Complete the sequence below:

40 50 60 ☐

K.CC.A.1

Complete the sequence below:

30 40 50 ☐

K.CC.A.1

5. Complete the sequence below:

60 70 80 ☐

K.CC.A.1

Complete the sequence below:

70 80 90 ☐

K.CC.A.1

6. Complete the sequence below:

20 30 40 ☐

K.CC.A.1

When you skip count, sometimes it helps to chant the numbers in a singsong voice.

WEEK 1 : DAY 5

ASSESSMENT

1. Complete the sequence below:

14 15 16 ☐

K.CC.A.1

2. Complete the sequence below:

50 60 70 ☐

K.CC.A.1

3. Complete the sequence below:

0 10 20 ☐

K.CC.A.1

4. Complete the sequence below:

0 1 2 ☐

K.CC.A

5. Complete the sequence below:

97 98 99 ☐

K.CC.A

6. Skip count from 0 to 30 by 5s.

☐ ☐ ☐ ☐
☐ ☐ ☐

K.CC.A

DAY 6
Challenge question

What number comes next in this sequence?

45 40 35 30 ☐

This week we are going to practice counting forward beginning from a given number within the known sequence.

You can find detailed video explanations of each problem in the book by visiting: ArgoPrep.com/ccmk

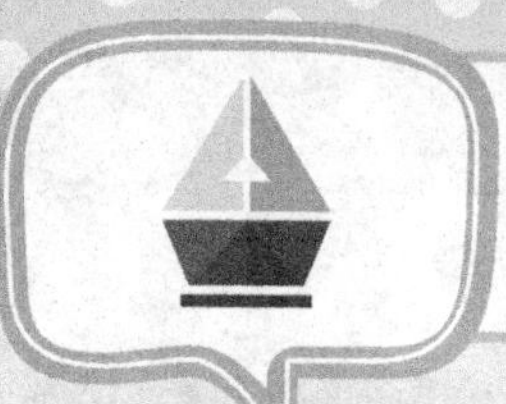

WEEK 2 : DAY 1

1. What number comes after 90?

   A. 91
   B. 92
   C. 93
   D. 94

K.CC.A.2

2. What number comes after 75?

   A. 74
   B. 75
   C. 76
   D. 77

K.CC.A.2

3. What number comes after 36?

   A. 34
   B. 35
   C. 36
   D. 37

K.CC.A.2

4. What number comes after 47?

   A. 47
   B. 48
   C. 49
   D. 50

K.CC.

5. What number comes after 16?

   A. 17
   B. 18
   C. 19
   D. 20

K.CC.A

6. What number comes after 65?

   A. 63
   B. 64
   C. 65
   D. 66

K.CC.A

It may be helpful to count starting at the closest 10 to find the number that comes next.

WEEK 2 : DAY 2

What number comes after 70?

A. 69
B. 70
C. 71
D. 72

K.CC.A.2

4. What number comes after 23?

A. 22
B. 23
C. 24
D. 25

K.CC.A.2

What number comes after 42?

A. 43
B. 44
C. 45
D. 46

K.CC.A.2

5. What number comes after 31?

A. 31
B. 32
C. 33
D. 34

K.CC.A.2

What number comes after 49?

A. 49
B. 50
C. 51
D. 52

K.CC.A.2

6. What number comes after 54?

A. 55
B. 56
C. 57
D. 58

K.CC.A.2

When we are looking at numbers that come after, we are really adding by 1.

WEEK 2 : DAY 3

1. What number comes after 52?

   A. 53
   B. 54
   C. 55
   D. 56

K.CC.A.2

2. What number comes after 25?

   A. 24
   B. 25
   C. 26
   D. 27

K.CC.A.2

3. What number comes after 17?

   A. 15
   B. 16
   C. 17
   D. 18

K.CC.A.2

4. What number comes after 3?

   A. 2
   B. 3
   C. 4
   D. 5

K.CC.A

5. What number comes after 83?

   A. 82
   B. 83
   C. 84
   D. 85

K.CC.A

6. What number comes after 62?

   A. 61
   B. 62
   C. 63
   D. 64

K.CC.A

It may be helpful to write out the numbers 1-10.

WEEK 2 : DAY 4

What number comes after 71?

A. 70
B. 71
C. 72
D. 73

K.CC.A.2

What number comes after 67?

A. 66
B. 67
C. 68
D. 69

K.CC.A.2

What number comes after 41?

A. 41
B. 42
C. 43
D. 44

K.CC.A.2

4. What number comes after 27?

A. 28
B. 29
C. 30
D. 31

K.CC.A.2

5. What number comes after 34?

A. 33
B. 34
C. 35
D. 36

K.CC.A.2

6. What number comes after 86?

A. 84
B. 85
C. 86
D. 87

K.CC.A.2

Remember, the numbers 0-9 repeat no matter what the value is in the 10s column.

WEEK 2 : DAY 5

ASSESSMENT

1. What number comes after 84?

   A. 83
   B. 84
   C. 85
   D. 86

K.CC.A.2

2. What number comes after 4?

   A. 5
   B. 6
   C. 7
   D. 8

K.CC.A.2

3. What number comes after 50?

   A. 50
   B. 51
   C. 52
   D. 53

K.CC.A.2

4. What number comes after 14?

   A. 15
   B. 16
   C. 17
   D. 19

K.CC.A

5. What number comes after 58?

   A. 56
   B. 57
   C. 58
   D. 59

K.CC.A

6. What number comes before 92?

   A. 90
   B. 91
   C. 92
   D. 93

K.CC.A

DAY 6 Challenge question

What number is missing?

46 ☐ 48 49

ARGOPREP.COM

VIDEO EXPLANATIONS

Week 3 is all about counting and writing numbers from 0 to 20. Let the fun begin!

You can find detailed video explanations of each problem in the book by visiting: ArgoPrep.com/ccmk

WEEK 3 : DAY I

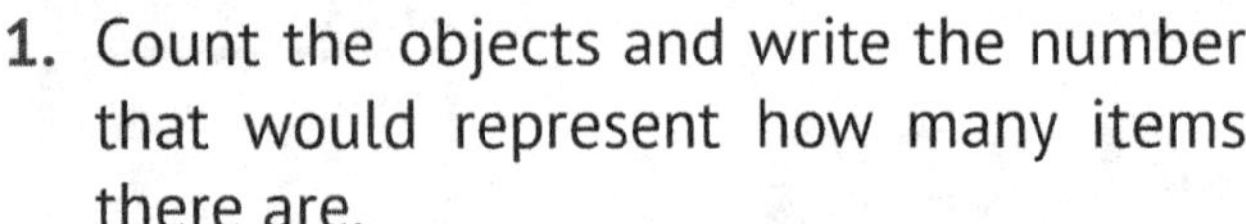

1. Count the objects and write the number that would represent how many items there are.

______________________ K.CC.A.3

2. Count the objects and write the number that would represent how many items there are.

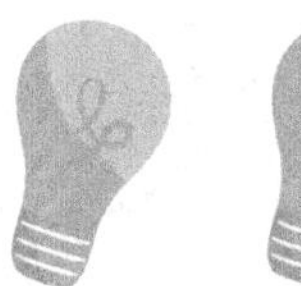

______________________ K.CC.A.3

3. Count the objects and write the number that would represent how many items there are.

______________________ K.CC.A.3

4. Count the objects and write the numb that would represent how many iter there are.

______________________ K.CC.A

5. Count the objects and write the numb that would represent how many iten there are.

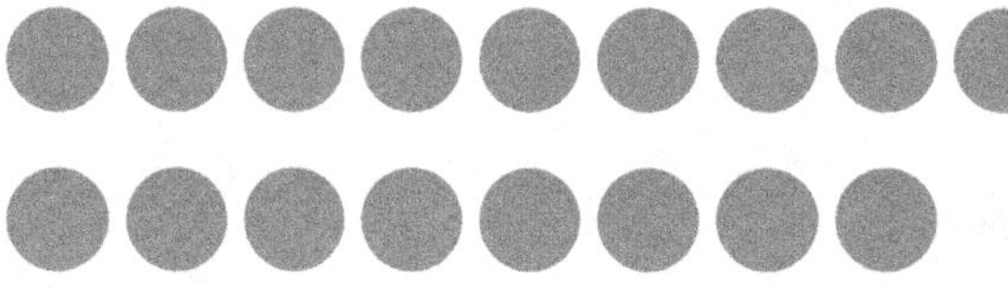

______________________ K.CC.A

6. Count the objects and write the numb that would represent how many iten there are.

______________________ K.CC.A

When counting, remember to point one finger for one object.

WEEK 3 : DAY 2

Count the objects and write the number that would represent how many items there are.

______________________ K.CC.A.3

4. Count the objects and write the number that would represent how many items there are.

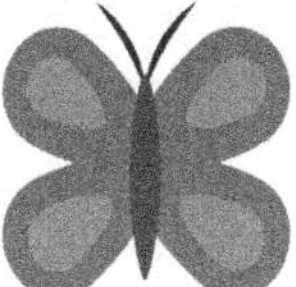

______________________ K.CC.A.3

Count the objects and write the number that would represent how many items there are.

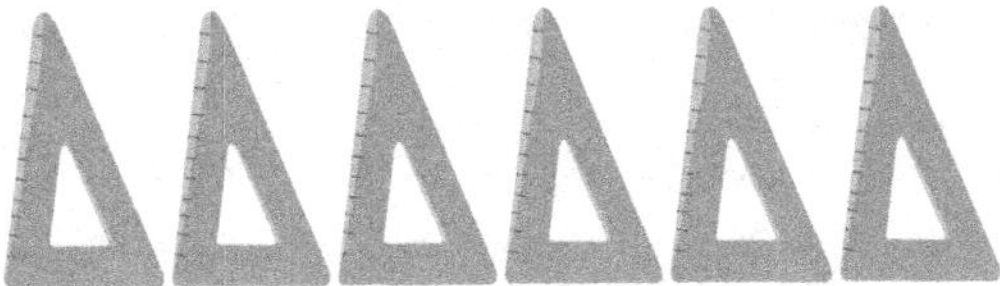

______________________ K.CC.A.3

5. Count the objects and write the number that would represent how many items there are.

______________________ K.CC.A.3

Count the objects and write the number that would represent how many items there are.

______________________ K.CC.A.3

6. Count the objects and write the number that would represent how many items there are.

______________________ K.CC.A.3

TIP of the DAY

As you count, sometimes it helps to cross out the objects as you say the numbers.

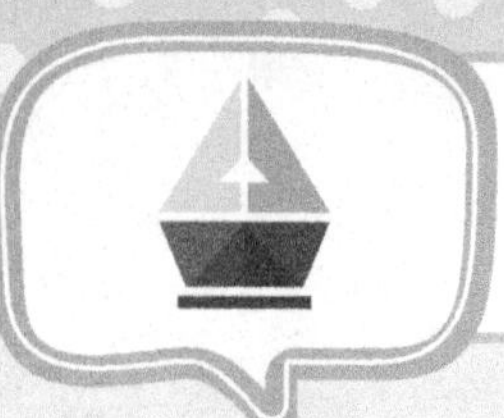

WEEK 3 : DAY 3

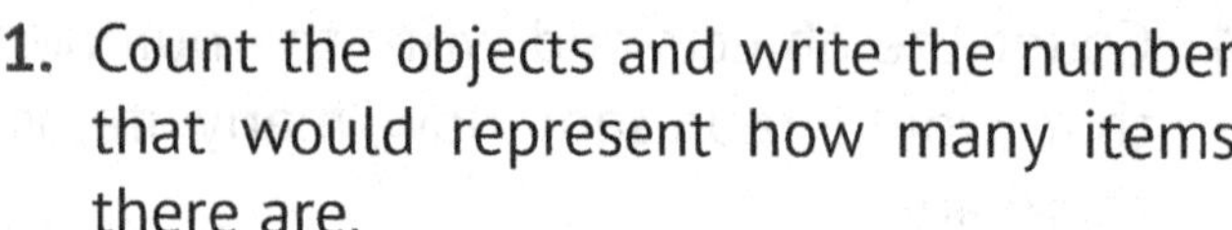

1. Count the objects and write the number that would represent how many items there are.

_______________ K.CC.A.3

2. Count the objects and write the number that would represent how many items there are.

_______________ K.CC.A.3

3. Count the objects and write the number that would represent how many items there are.

_______________ K.CC.A.3

4. Count the objects and write the numb that would represent how many iter there are.

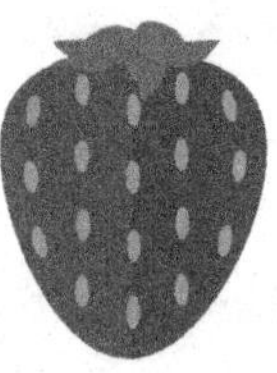

_______________ K.CC.A

5. Count the objects and write the numb that would represent how many iter there are.

_______________ K.CC.A

6. Count the objects and write the numb that would represent how many iten there are.

_______________ K.CC.A

When determining how many objects, it might be helpful to group objects together.

WEEK 3 : DAY 4

Write the numeral represented by the word "three"

K.CC.A.3

Write the numeral represented by the word "seven"

K.CC.A.3

Write the numeral represented by the word "five"

K.CC.A.3

4. Write the numeral represented by the word "one"

K.CC.A.3

5. Write the numeral represented by the word "nine"

K.CC.A.3

6. Write the numeral represented by the word "eleven"

K.CC.A.3

If you are having trouble, start with the easier questions first and then work on the hardest questions last.

WEEK 3 : DAY 5

ASSESSMENT

1. Write the numeral represented by the word "ten"

   ____

   K.CC.A.3

2. Write the numeral represented by the word "four"

   ____

   K.CC.A.3

3. Write the numeral represented by the word "eight"

   ____

   K.CC.A.3

4. Write the numeral represented by the word "two"

   ____

   K.CC.A

5. Write the numeral represented by the word "six"

   ____

   K.CC.A

6. Write the numeral represented by the word "twenty-seven"

   ____

   K.CC.A

DAY 6
Challenge question

What word represents the number 82?

ARGOPREP.COM

VIDEO
EXPLANATIONS

How well do you know your numbers? We will practice counting objects but also saying the number names.

You can find detailed video explanations of each problem in the book by visiting: ArgoPrep.com/ccmk

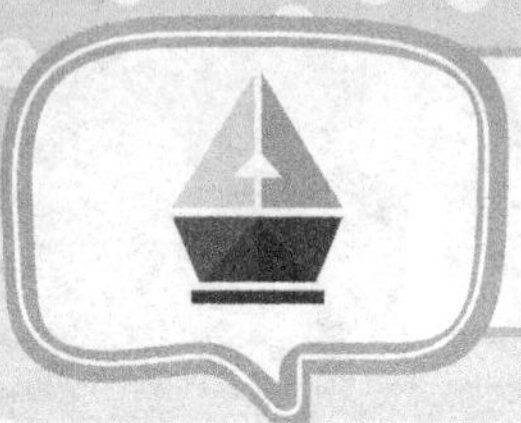

WEEK 4 : DAY 1

1. Which word represents the number 13?

   A. Twelve
   B. Thirteen
   C. Fourteen
   D. Fifteen

   K.CC.B.4.a

2. Which word represents the number 4?

   A. Four
   B. Five
   C. Six
   D. Seven

   K.CC.B.4.a

3. Which word represents the number 10?

   A. Eight
   B. Nine
   C. Ten
   D. Eleven

   K.CC.B.4.a

4. Which word represents the number 1?

   A. One
   B. Two
   C. Three
   D. Four

   K.CC.B.4

5. Which word represents the number 17?

   A. Fifteen
   B. Sixteen
   C. Seventeen
   D. Eighteen

   K.CC.B.4

6. Which word represents the number 6?

   A. Three
   B. Four
   C. Five
   D. Six

   K.CC.B.4

Sound out words if you are unsure which number is represented by which word.

WEEK 4 : DAY 2

How many objects are there?

A. 1
B. 2
C. 3
D. 4

K.CC.B.4.a

4. How many objects are there?

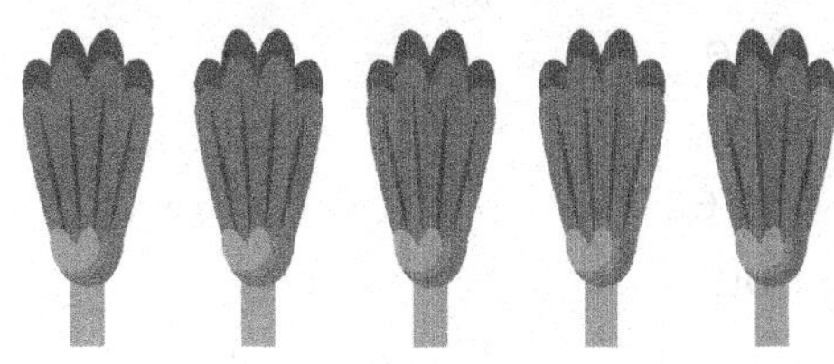

A. 2
B. 3
C. 4
D. 5

K.CC.B.4.a

How many objects are there?

A. 6
B. 7
C. 8
D. 9

K.CC.B.4.a

5. How many objects are there?

A. 6
B. 7
C. 8
D. 9

K.CC.B.4.a

How many objects are there?

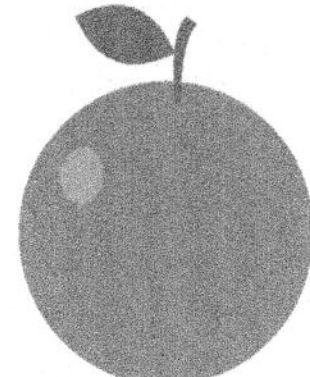

A. 1
B. 2
C. 3
D. 4

K.CC.B.4.a

6. How many objects are there?

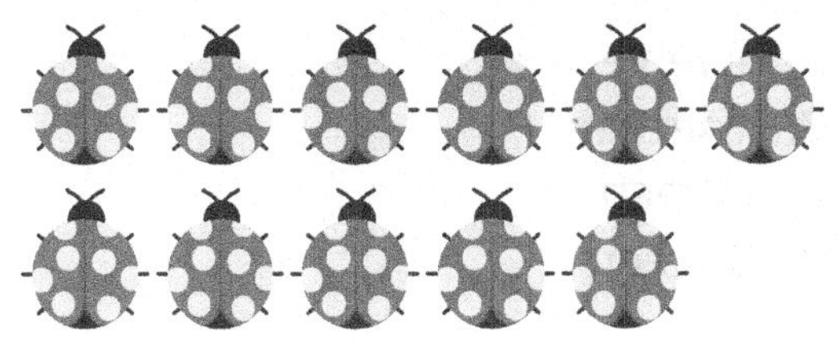

A. 9
B. 10
C. 11
D. 12

K.CC.B.4.a

TIP of the DAY

When completing multiple choice, always check your chosen answer against the other choices.

WEEK 4 : DAY 3

1. Which word represents the number 2?

   A. One
   B. Two
   C. Three
   D. Four

K.CC.B.4.a

2. Which word represents the number 14?

   A. Eleven
   B. Thirteen
   C. Fourteen
   D. Fifteen

K.CC.B.4.a

3. Which word represents the number 16?

   A. Twelve
   B. Thirteen
   C. Fifteen
   D. Sixteen

K.CC.B.4.a

4. Which word represents the number 8?

   A. Six
   B. Seven
   C. Eight
   D. Nine

K.CC.B.4

5. Which word represents the number 11

   A. Ten
   B. Eleven
   C. Twelve
   D. Thirteen

K.CC.B.4

6. Which word represents the number 5?

   A. Five
   B. Six
   C. Seven
   D. Eight

K.CC.B.4

If you cannot chose an answer, it can be helpful to eliminate choices you know are wrong.

WEEK 4 : DAY 4

How many objects are there?

_______________ K.CC.B.4.a

4. How many objects are there?

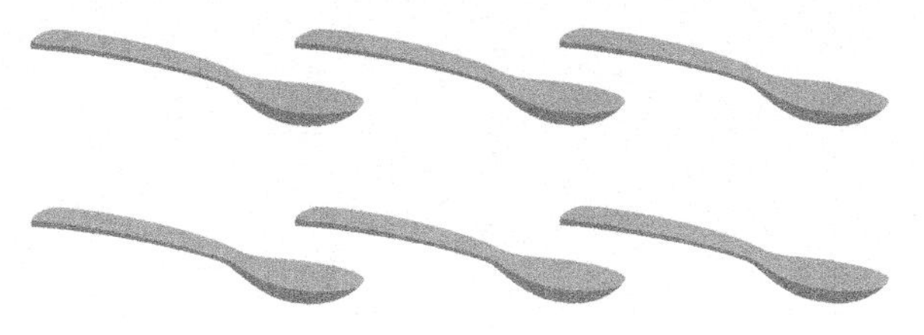

_______________ K.CC.B.4.a

How many objects are there?

_______________ K.CC.B.4.a

5. How many objects are there?

_______________ K.CC.B.4.a

How many objects are there?

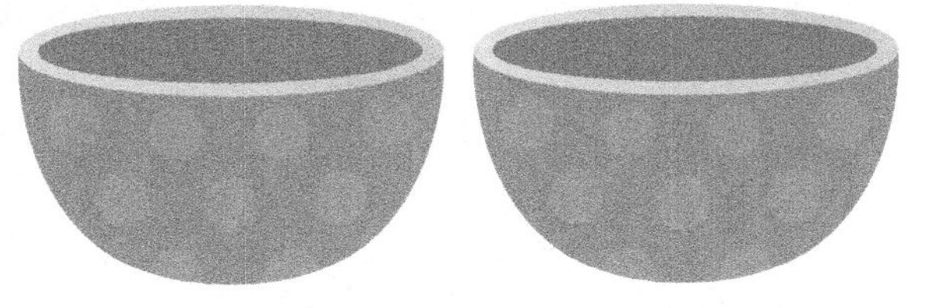

_______________ K.CC.B.4.a

6. How many objects are there?

_______________ K.CC.B.4.a

When finding the missing number in a subtraction problem, think of using a bar model to find the answer.

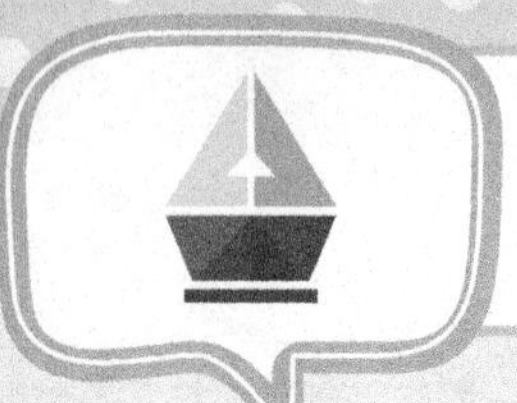

WEEK 4 : DAY 5

ASSESSMENT

1. What word represents the number 7?

____

K.CC.B.4.a

2. What word represents the number 15?

____

K.CC.B.4.a

3. What word represents the number 9?

____

K.CC.B.4.a

4. What word represents the number 3?

____

K.CC.B.4

5. What word represents the number 12?

____

K.CC.B.4

6. Draw a set of 17 objects.

K.CC.B.4

DAY 6
Challenge question

What number is represented by one hundred six?

1.OA.B.4

What is one more than four? In week 5, we will be working with the concept of "one more".

You can find detailed video explanations of each problem in the book by visiting:
ArgoPrep.com/ccmk

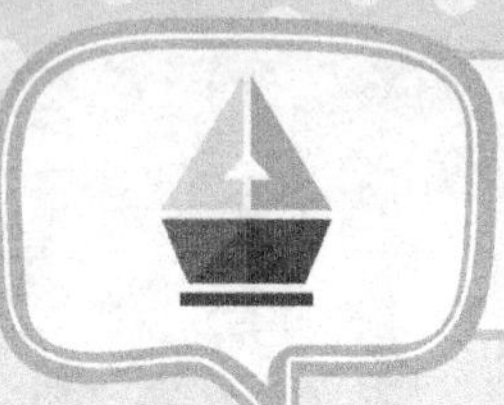

WEEK 5 : DAY 1

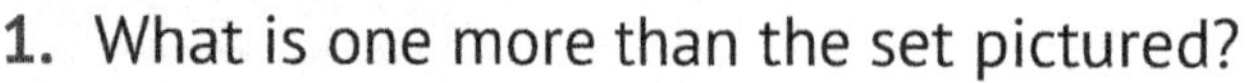

1. What is one more than the set pictured?

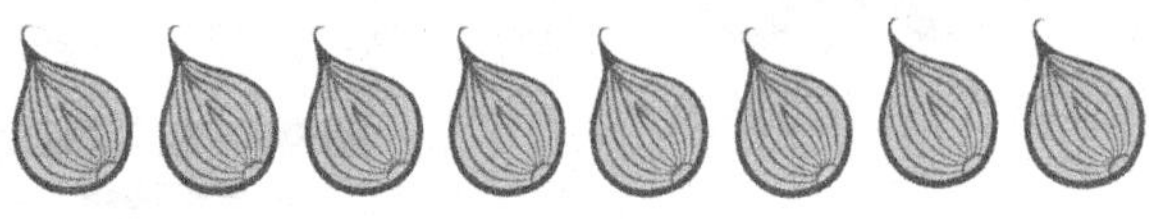

A. 7
B. 8
C. 9
D. 10

K.CC.B.4.b

2. What is one more than the set pictured?

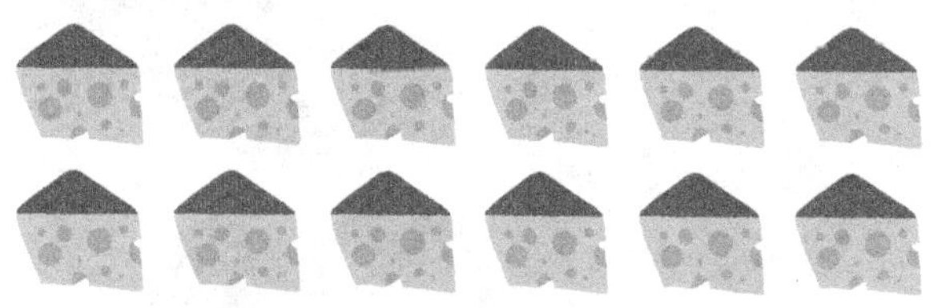

A. 10
B. 11
C. 12
D. 13

K.CC.B.4.b

3. What is one more than the set pictured?

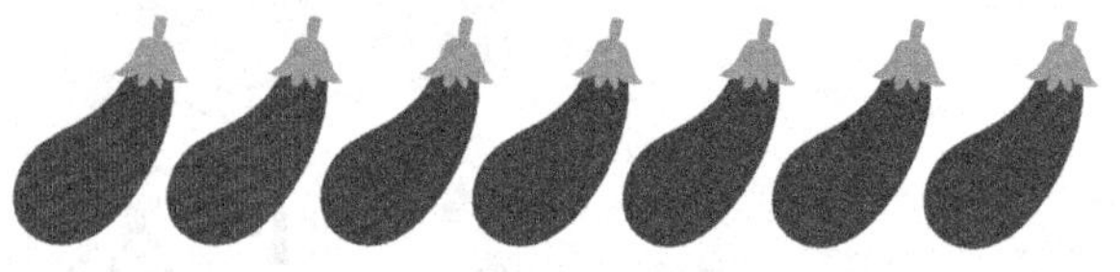

A. 6
B. 7
C. 8
D. 9

K.CC.B.4.b

4. What is one more than the set picturec

A. 2
B. 3
C. 4
D. 5

K.CC.B.4

5. What is one more than the set picturec

A. 2
B. 3
C. 4
D. 5

K.CC.B.4

6. What is one more than the set picturec

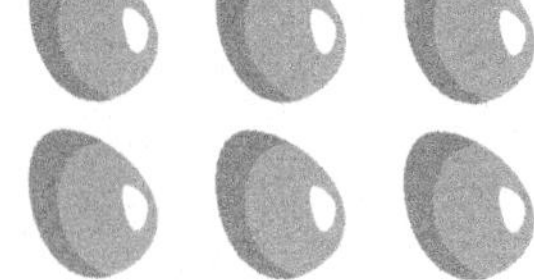

A. 8
B. 9
C. 10
D. 11

K.CC.B.4

TIP of the DAY

It may help to draw a model that represents the question.

WEEK 5 : DAY 2

What number comes after six?

____

K.CC.B.4.b

4. What number comes after seventeen?

____

K.CC.B.4.b

What number comes after thirteen?

____

K.CC.B.4.b

5. What number comes after one?

____

K.CC.B.4.b

What number comes after ten?

____

K.CC.B.4.b

6. What number comes after eleven?

____

K.CC.B.4.b

You should always have extra paper available to work out problems.

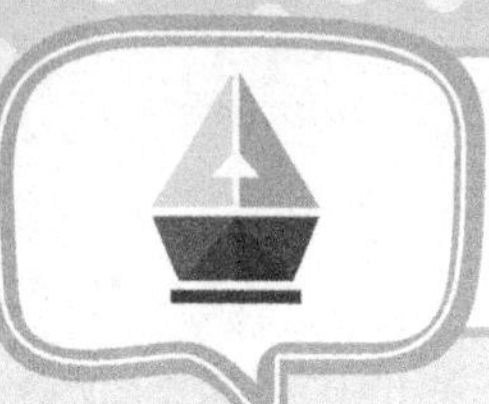

WEEK 5 : DAY 3

1. What number comes after eight?

   A. 9
   B. 10
   C. 11
   D. 12

K.CC.B.4.b

2. What number comes after nineteen?

   A. 20
   B. 19
   C. 18
   D. 17

K.CC.B.4.b

3. What number comes after fourteen?

   A. 14
   B. 15
   C. 16
   D. 17

K.CC.B.4.b

4. What number comes after two?

   A. 2
   B. 3
   C. 4
   D. 5

K.CC.B.4

5. What number comes after four?

   A. 5
   B. 6
   C. 7
   D. 8

K.CC.B.4

6. What number comes after nine?

   A. 12
   B. 11
   C. 10
   D. 9

K.CC.B.4

A number line might be useful to solve these problems.

WEEK 5 : DAY 4

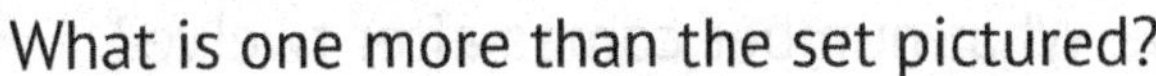

What is one more than the set pictured?

______________________ K.CC.B.4.b

What is one more than the set pictured?

______________________ K.CC.B.4.b

What is one more than the set pictured?

______________________ K.CC.B.4.b

4. What is one more than the set pictured?

______________________ K.CC.B.4.b

5. What is one more than the set pictured?

______________________ K.CC.B.4.b

6. What is one more than the set pictured?

______________________ K.CC.B.4.b

TIP of the DAY

If you can't solve it mentally, draw a picture. For example, 9 + 3 = 10 + 2 = 12.

WEEK 5 : DAY 5

ASSESSMENT

1. What number comes after five?

   A. 5
   B. 6
   C. 7
   D. 8

   K.CC.B.4.b

2. What number comes after eighteen?

   A. 16
   B. 17
   C. 18
   D. 19

   K.CC.B.4.b

3. What number comes after three?

   A. 1
   B. 2
   C. 3
   D. 4

   K.CC.B.4.b

4. What number comes after twelve?

   A. 10
   B. 11
   C. 12
   D. 13

   K.CC.B.4

5. What number comes after fifteen?

   A. 15
   B. 16
   C. 17
   D. 18

   K.CC.B.4

6. What number comes after fifty-nine?

   A. 57
   B. 58
   C. 59
   D. 60

   K.CC.B.4

DAY 6
Challenge question

What number comes three steps after thirty-five?

1.OA.C.5

This week continues on focusing on the standard where we need to understand that each successive number name refers to a quantity that is one larger.

You can find detailed video explanations of each problem in the book by visiting: ArgoPrep.com/ccmk

WEEK 6 : DAY 1

1. What number is one larger than 6?

____

K.CC.B.4.c

2. What number is one larger than 19?

____

K.CC.B.4.c

3. What number is one larger than 2?

____

K.CC.B.4.c

4. What number is one larger than 5?

____

K.CC.B.4

5. What number is one larger than 16?

____

K.CC.B.4

6. What number is one larger than 10?

____

K.CC.B.4

It may be helpful to practice counting to 100 before you complete these problems.

WEEK 6 : DAY 2

What number is one larger than 11?

____

K.CC.B.4.c

4. What number is one larger than 18?

____

K.CC.B.4.c

What number is one larger than 15?

____

K.CC.B.4.c

5. What number is one larger than 7?

____

K.CC.B.4.c

What number is one larger than 1?

____

K.CC.B.4.c

6. What number is one larger than 10?

____

K.CC.B.4.c

If you write out the numbers 1-100, you can use them to solve these problems.

WEEK 6 : DAY 3

1. What number is one larger than 1?

____

K.CC.B.4.c

4. What number is one larger than 15?

____

K.CC.B.4

2. What number is one larger than 12?

____

K.CC.B.4.c

5. What number is one larger than 3?

____

K.CC.B.4

3. What number is one larger than 17?

____

K.CC.B.4.c

6. What number is one larger than 13?

____

K.CC.B.4

If you are having trouble with these problems, practice counting with objects such as pennies or beans.

WEEK 6 : DAY 4

What number is one larger than thirteen?

A. 14
B. 15
C. 16
D. 17

K.CC.B.4.c

What number is one larger than eighteen?

A. 17
B. 18
C. 19
D. 20

K.CC.B.4.c

What number is one larger than four?

A. 4
B. 5
C. 6
D. 7

K.CC.B.4.c

4. What number is one larger than eleven?

A. 10
B. 11
C. 12
D. 13

K.CC.B.4.c

5. What number is one larger than eight?

A. 7
B. 8
C. 9
D. 10

K.CC.B.4.c

6. What number is one larger than six?

A. 4
B. 5
C. 6
D. 7

K.CC.B.4.c

Remember to start with the number mentioned in the problem.

WEEK 6 : DAY 5

ASSESSMENT

1. What number is one larger than fourteen?

   A. 14
   B. 15
   C. 16
   D. 17

   K.CC.B.4.c

2. What number is one larger than nineteen?

   A. 18
   B. 19
   C. 20
   D. 21

   K.CC.B.4.c

3. What number is one larger than five?

   A. 4
   B. 5
   C. 6
   D. 7

   K.CC.B.4.c

4. What number is one larger than nine?

   A. 10
   B. 11
   C. 12
   D. 13

   K.CC.B.4

5. What number is one larger than sixteen

   A. 15
   B. 16
   C. 17
   D. 18

   K.CC.B.4

6. What number is one larger than one hundred ten?

   A. 11
   B. 111
   C. 10
   D. 112

   K.CC.B.4

DAY 6 Challenge question

What number is ten less than eighty-two?

1.OA.C.6

Week 7 is all about counting to answer "how many" questions.

You can find detailed video explanations of each problem in the book by visiting: ArgoPrep.com/ccmk

WEEK 7 : DAY 1

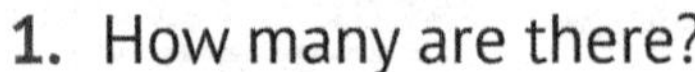

1. How many are there?

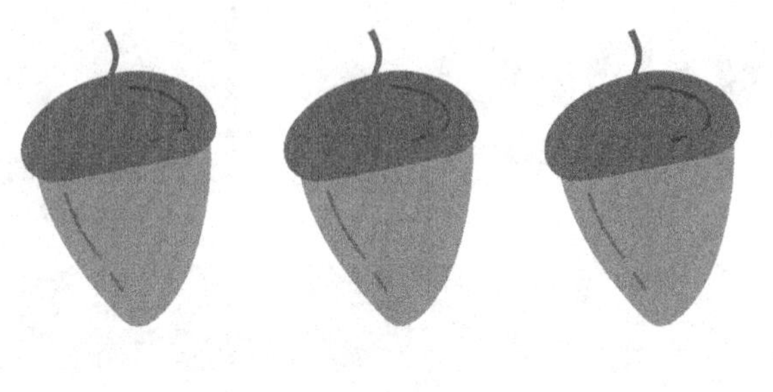

______________ K.CC.B.5

2. How many are there?

______________ K.CC.B.5

3. How many are there?

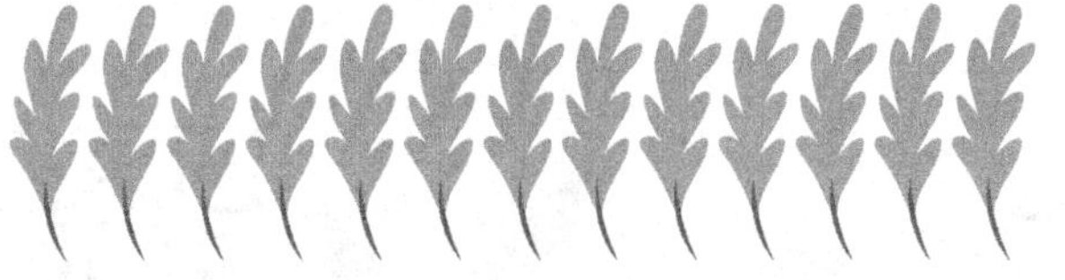

______________ K.CC.B.4.c

4. How many are there?

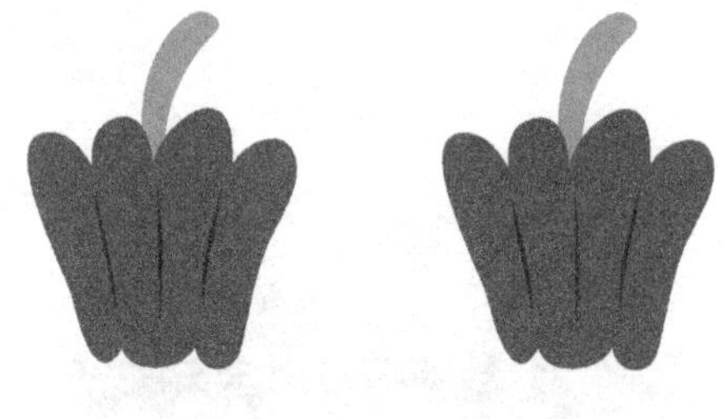

______________ K.CC.B

5. How many are there?

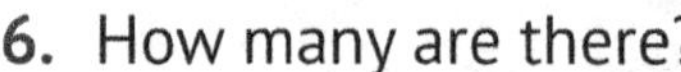

______________ K.CC.B

6. How many are there?

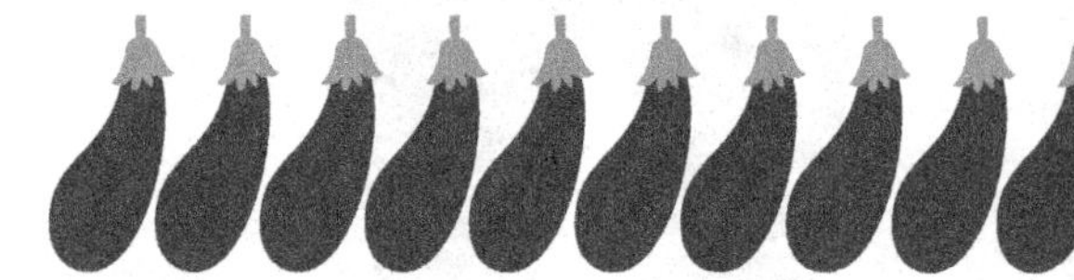

______________ K.CC.B

TIP of the DAY

The type of object does not impact the number of objects that are shown in a diagram.

WEEK 7 : DAY 2

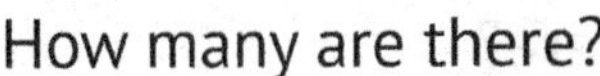

How many are there?

______________________ K.CC.B.5

4. How many are there?

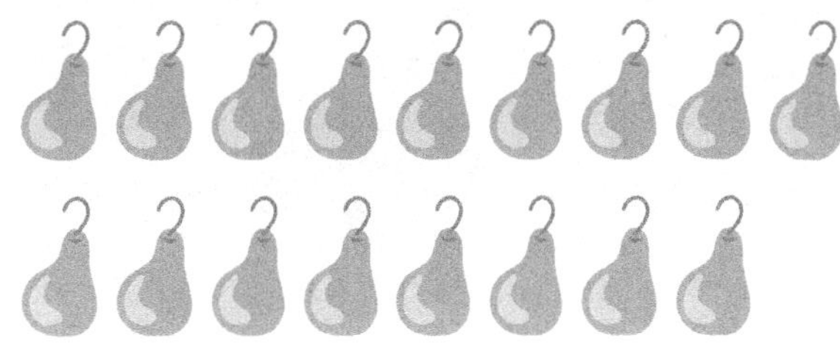

______________________ K.CC.B.5

How many are there?

______________________ K.CC.B.5

5. How many are there?

______________________ K.CC.B.5

How many are there?

______________________ K.CC.B.5

6. How many are there?

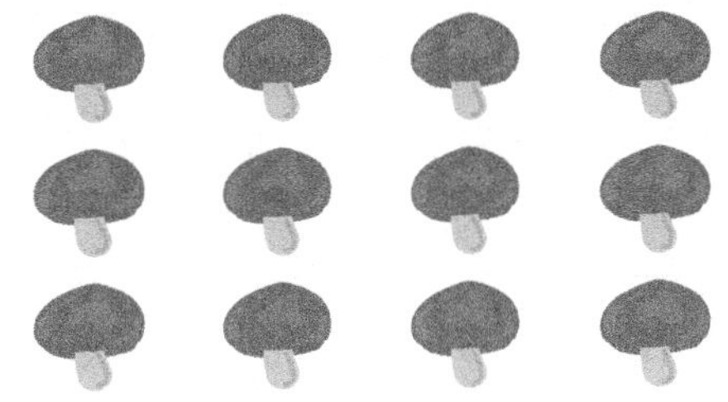

______________________ K.CC.B.5

When counting large groups, use a pencil to mark each object as you count.

WEEK 7 : DAY 3

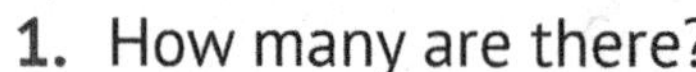

1. How many are there?

_______________ K.CC.B.5

2. How many are there?

_______________ K.CC.B.5

3. How many are there?

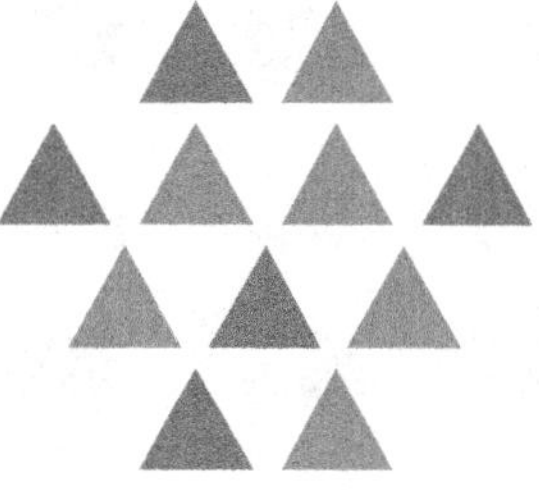

_______________ K.CC.B.5

4. How many are there?

_______________ K.CC.B

5. How many are there?

_______________ K.CC.B

6. How many are there?

_______________ K.CC.B.

TIP of the DAY

The color and shape of the group does not have an impact on how many objects are in the group.

WEEK 7 : DAY 4

Draw 4 objects.

K.CC.B.5

4. Draw 2 objects.

K.CC.B.5

Draw 10 objects.

K.CC.B.5

5. Draw 11 objects.

K.CC.B.5

Draw 6 objects.

K.CC.B.5

6. Draw 8 objects.

K.CC.B.5

As you draw each object, make sure it is small enough to fit all the others that need to be drawn.

WEEK 7 : DAY 5

ASSESSMENT

1. Draw 7 objects.

K.CC.B.5

2. Draw 3 objects.

K.CC.B.5

3. Draw 12 objects.

K.CC.B.5

4. Draw 9 objects.

K.CC.B

5. Draw 5 objects.

K.CC.B

6. Draw 13 objects.

K.CC.B

DAY 6
Challenge question

Count the number of objects you drew on this page. How many objects did you draw in all?

1.OA.D.7

In week 8 we will identify whether the number of objects in one group is greater than, less than, or equal to the number of objects in another group.

You can find detailed video explanations of each problem in the book by visiting: ArgoPrep.com/ccmk

WEEK 8 : DAY 1

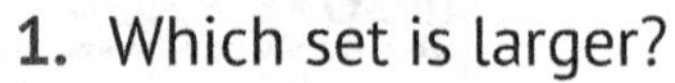

1. Which set is larger?

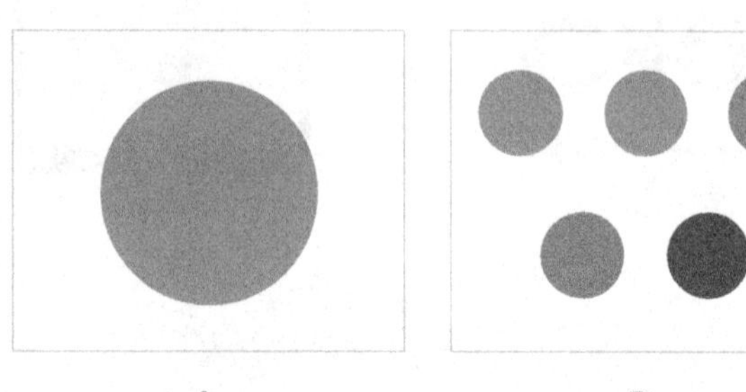

A. **B.**

K.CC.C.6

2. Which set is larger?

A. **B.**

K.CC.C.6

3. Which set is larger?

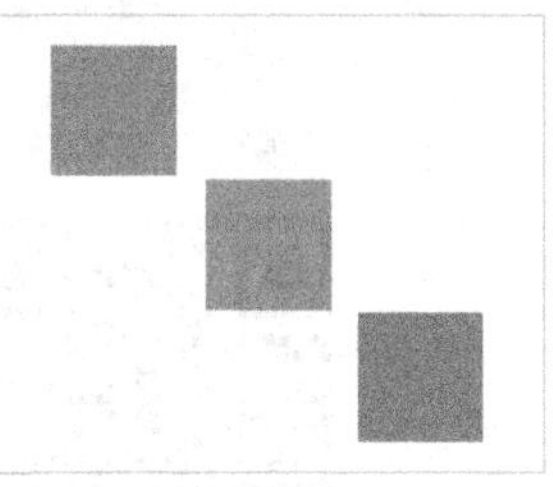 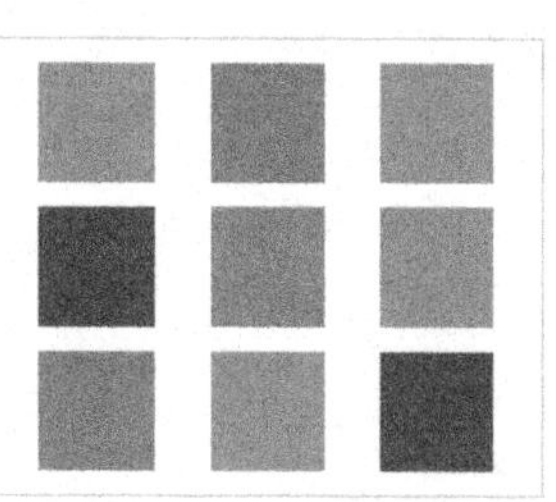

A. **B.**

K.CC.C.6

4. Which set is larger?

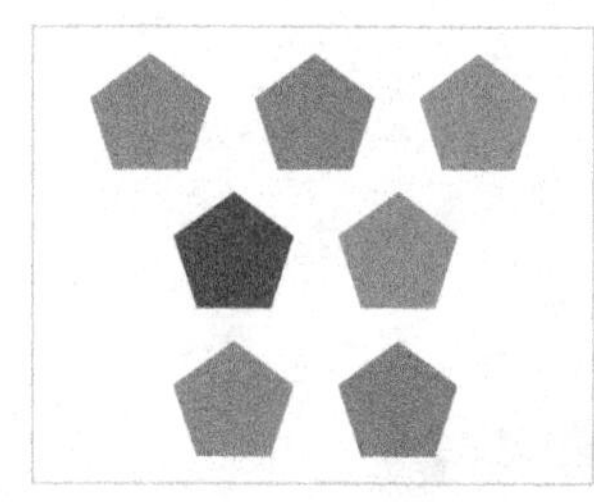 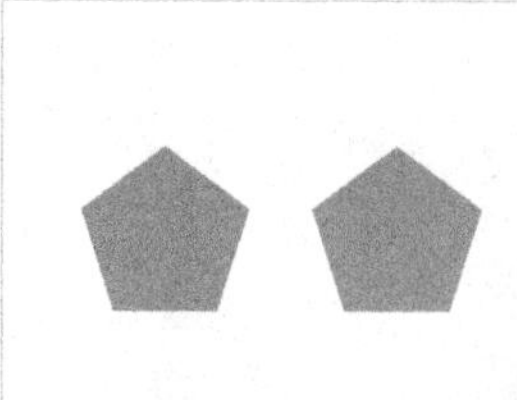

A. **B.**

K.CC.C

5. Which set is larger?

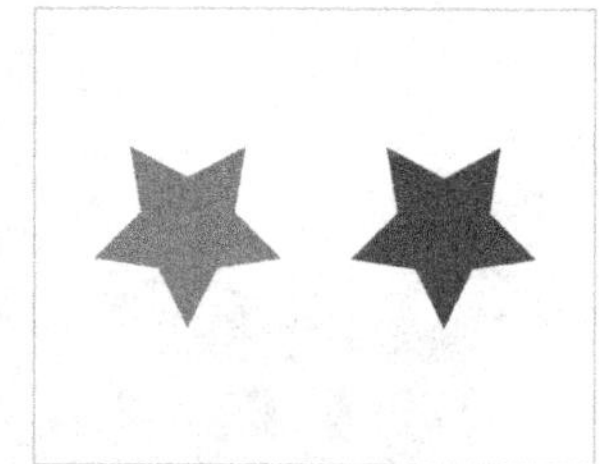 

A. **B.**

K.CC.C

6. Which set is larger?

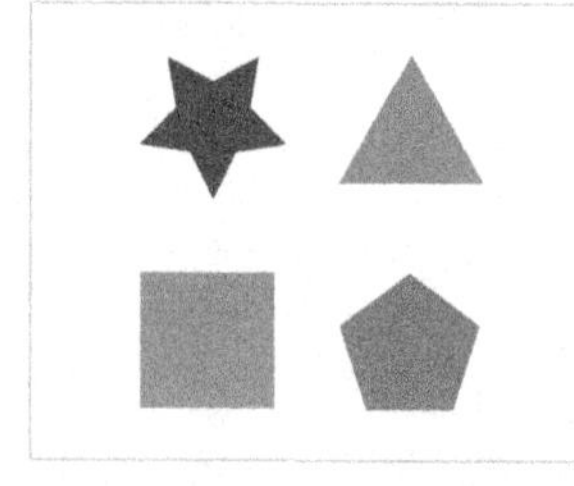 

A. **B.**

K.CC.C

The size of the objects does not impact which set is larger.

WEEK 8 : DAY 2

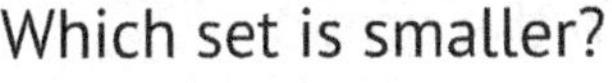

Which set is smaller?

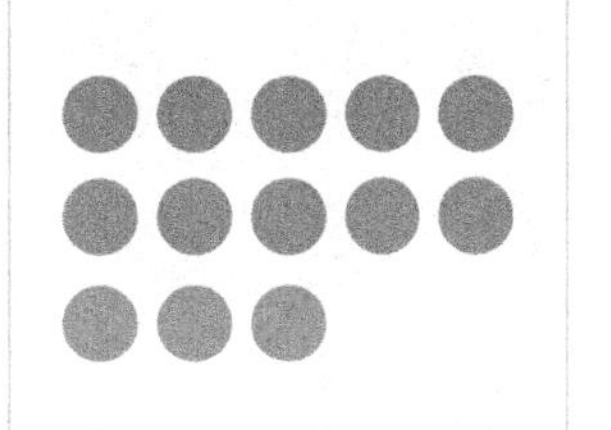

A. B.

K.CC.C.6

4. Which set is smaller?

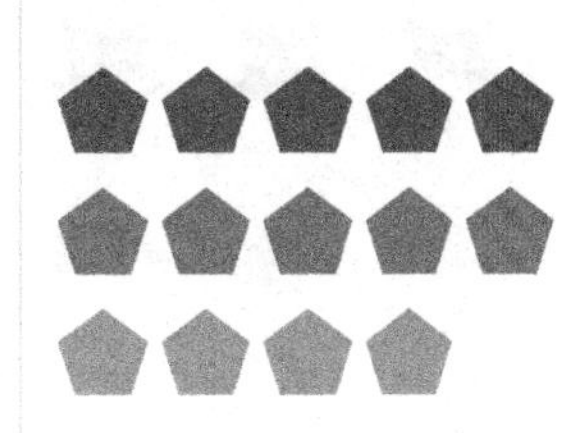

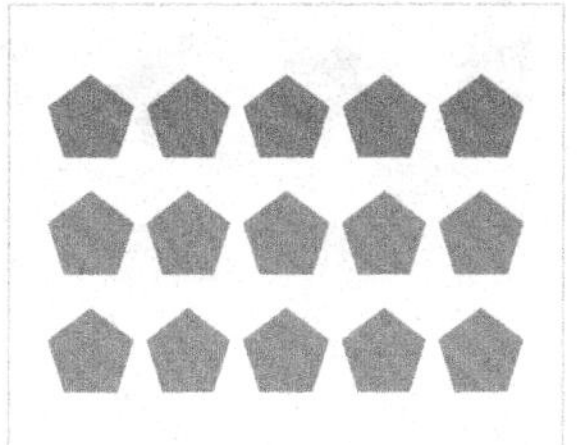

A. B.

K.CC.C.6

Which set is smaller?

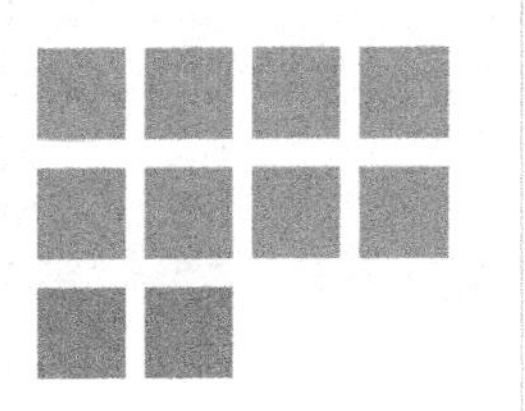

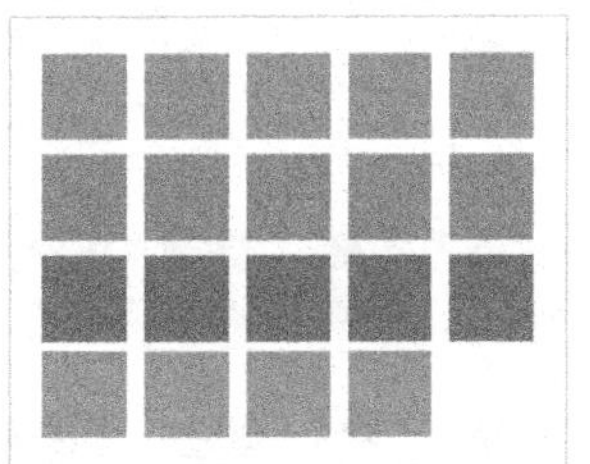

A. B.

K.CC.C.6

5. Which set is smaller?

A. B.

K.CC.C.6

Which set is smaller?

A. B.

K.CC.C.6

6. Which set is smaller?

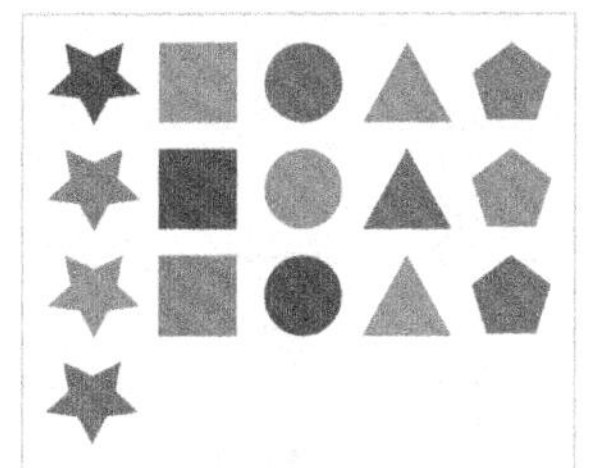

A. B.

K.CC.C.6

TIP of the DAY

The type of the objects does not impact which set is larger.

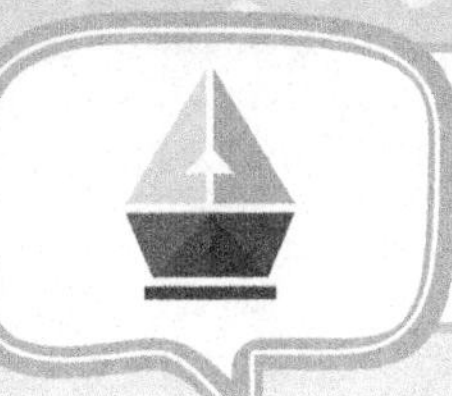

WEEK 8 : DAY 3

1. Circle 2 objects in the group.

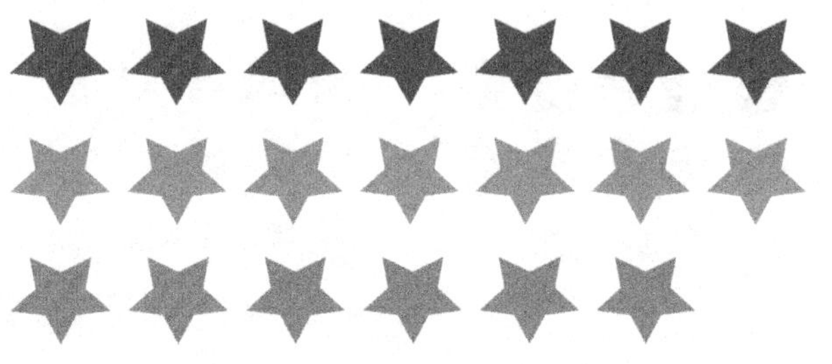

K.CC.C.6

4. Circle 5 objects in the group.

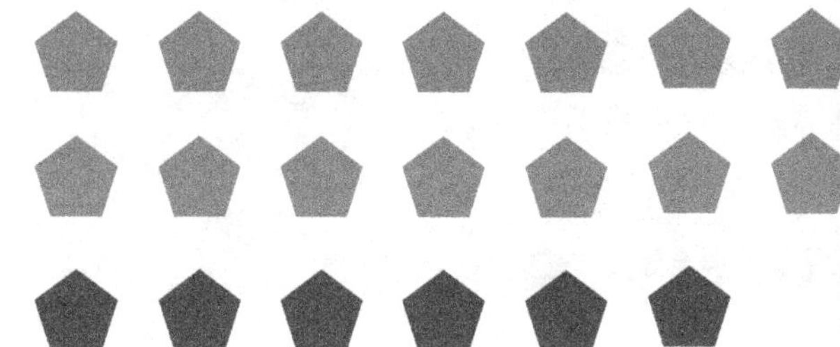

K.CC.C

2. Circle 13 objects in the group.

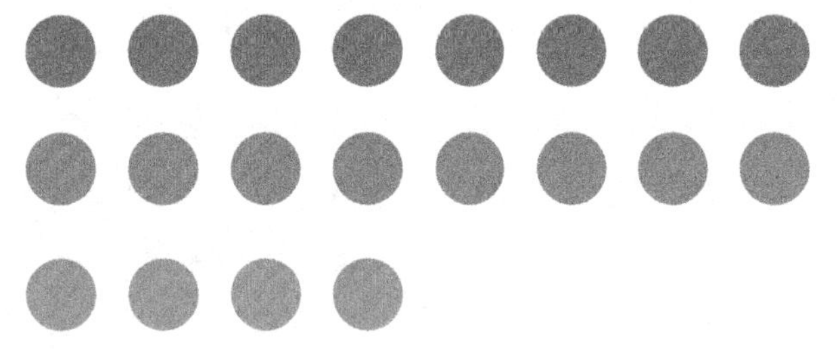

K.CC.C.6

5. Circle 16 objects in the group.

K.CC.C

3. Circle 7 objects in the group.

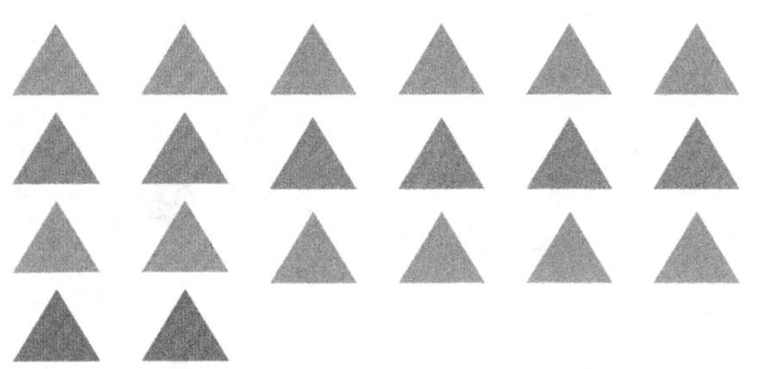

K.CC.C.6

6. Circle 10 objects in the group.

K.CC.C

TIP of the DAY

Be sure to count the group you create after you circle them.

WEEK 8 : DAY 4

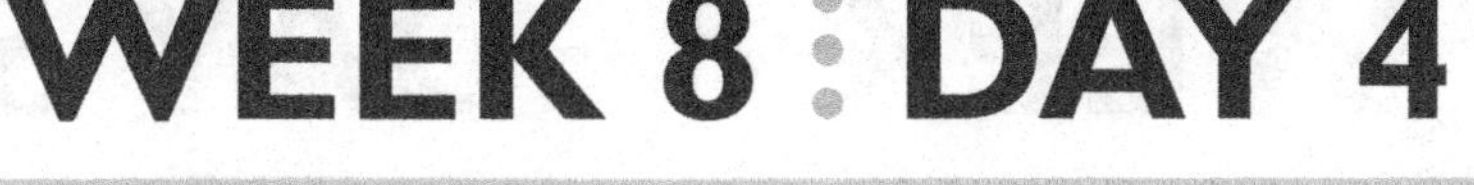
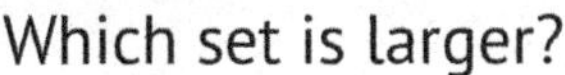

Which set is larger?

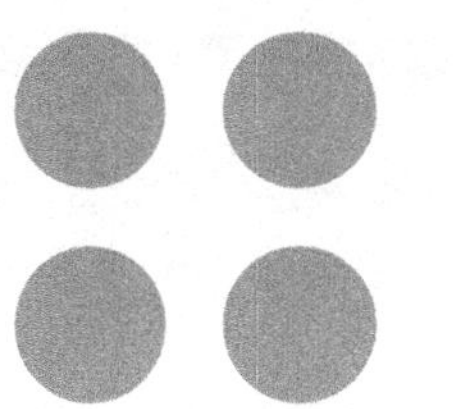
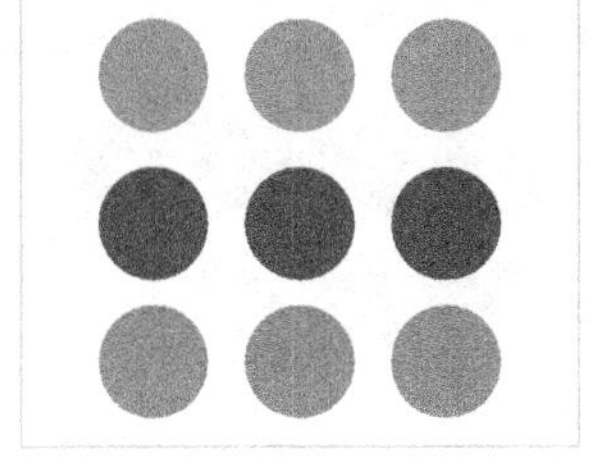

A. B. K.CC.C.6

4. Which set is smaller?

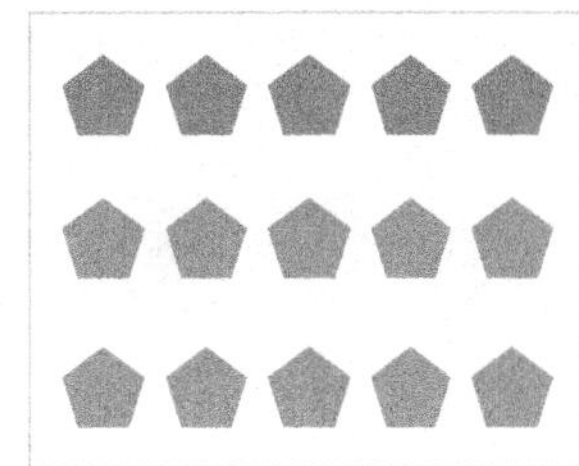
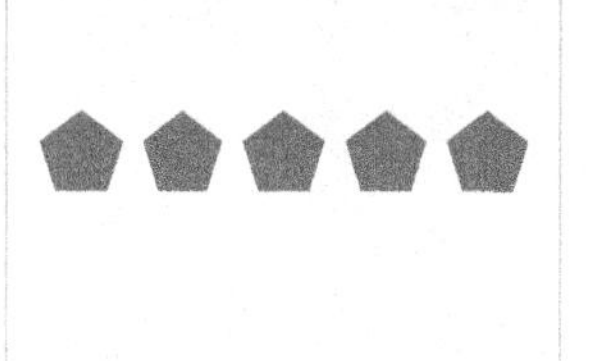

A. B. K.CC.C.6

Which set is smaller?

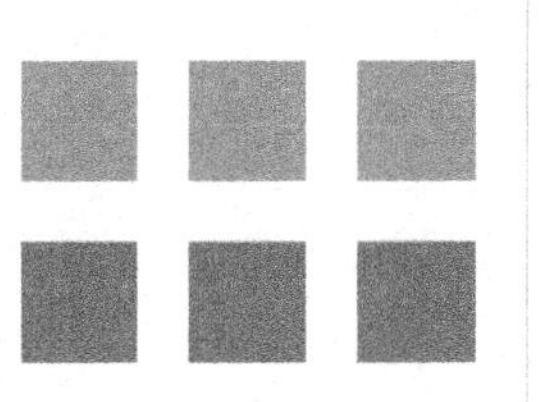
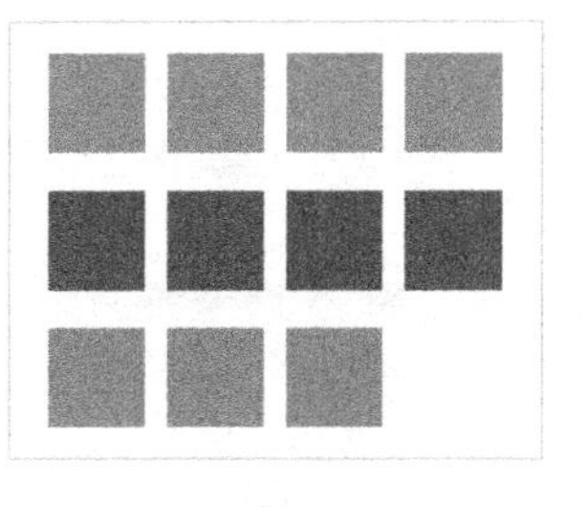
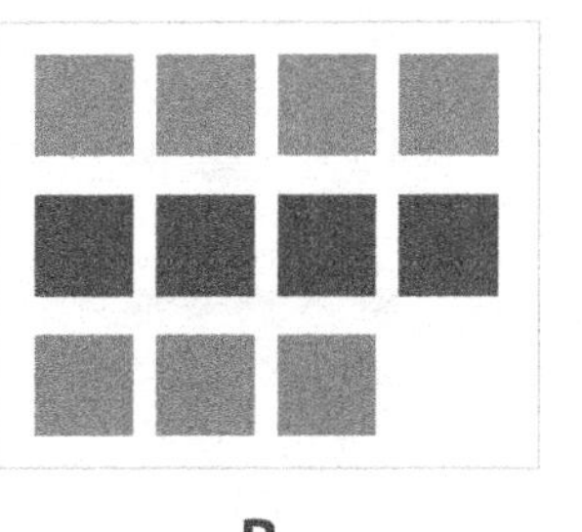

A. B. K.CC.C.6

5. Which set is smaller?

A. B. K.CC.C.6

Which set is larger?

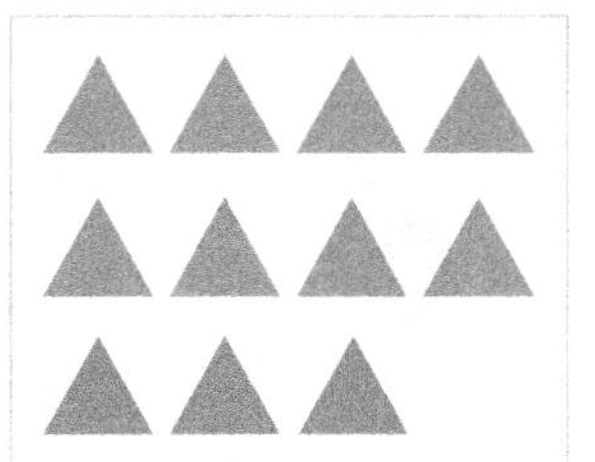

A. B. K.CC.C.6

6. Which set is larger?

A. B. K.CC.C.6

TIP of the DAY

It may be helpful to write the number that represents how many objects are in the group.

WEEK 8 : DAY 5

ASSESSMENT

1. Which set is smaller?

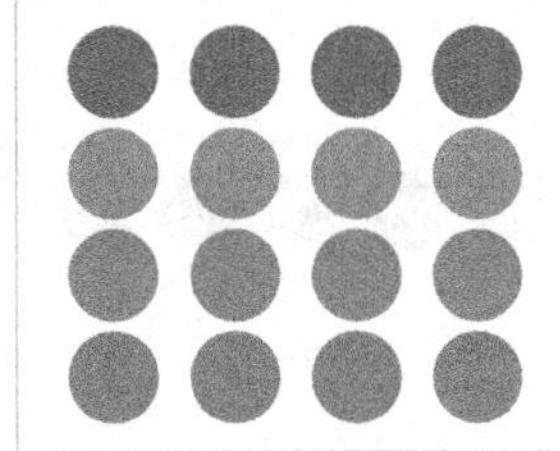

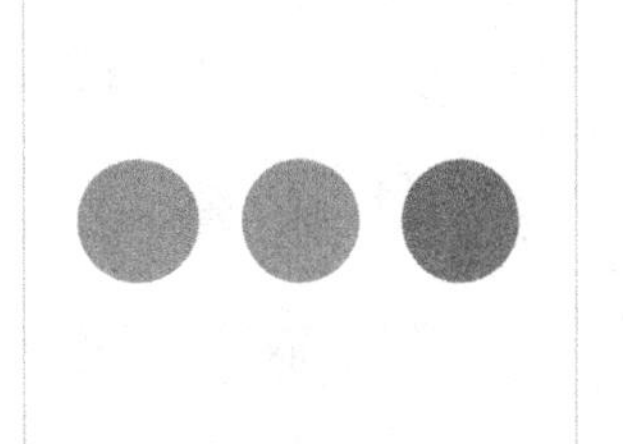

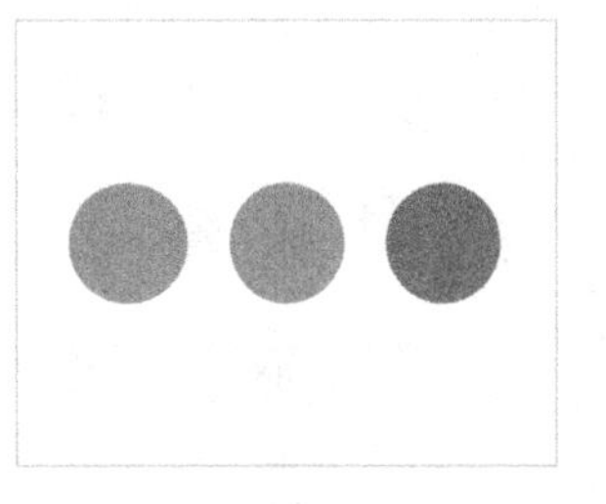

A. **B.**

K.CC.C.6

2. Which set is larger?

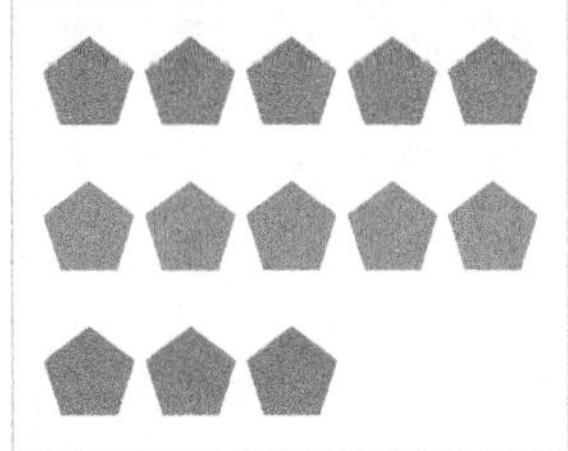

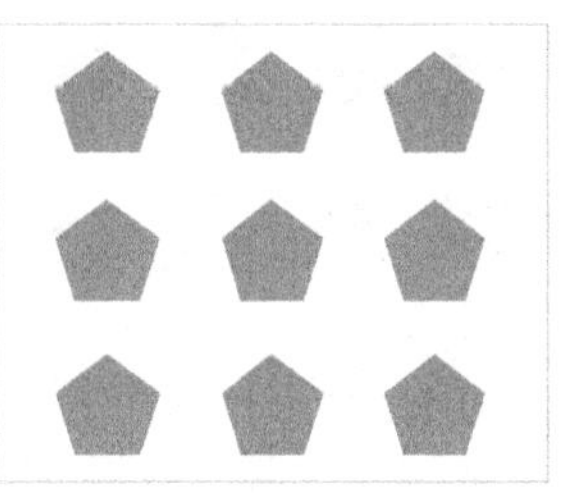

A. **B.**

K.CC.C.6

3. Which set is smaller?

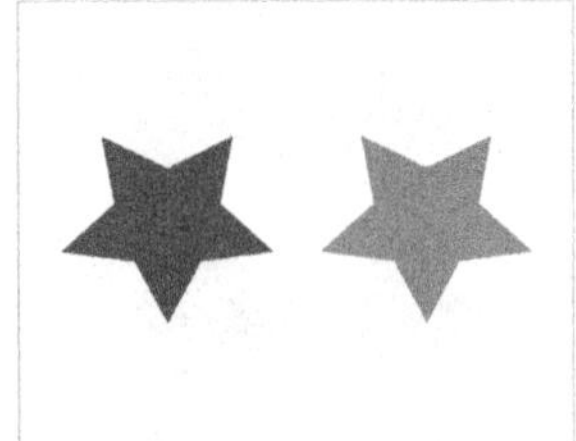

A. **B.**

K.CC.C.6

4. Which set is larger?

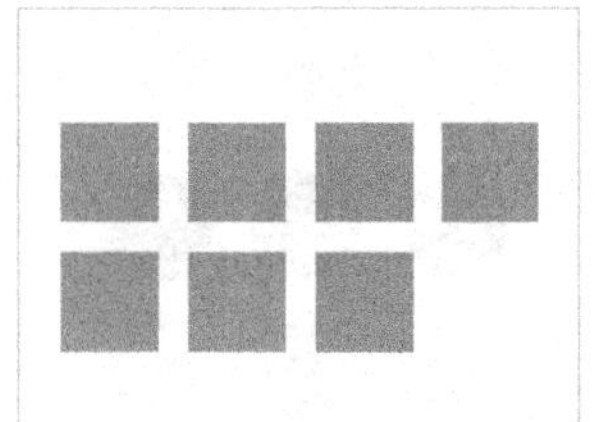

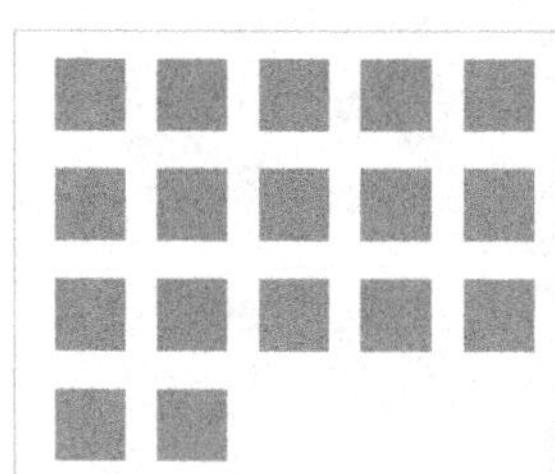

A. **B.**

K.CC.C

5. Which set is larger?

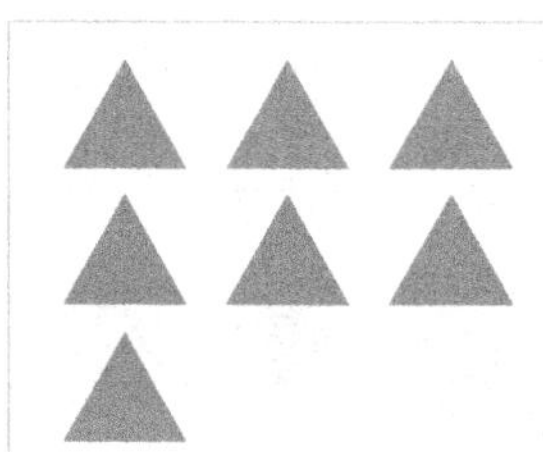

A. **B.**

K.CC.C

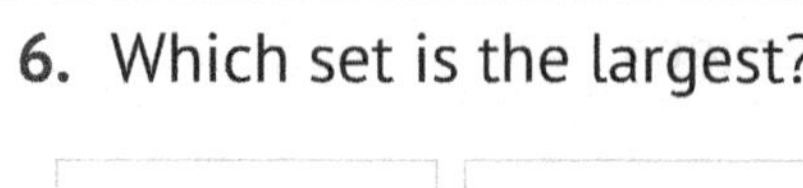

6. Which set is the largest?

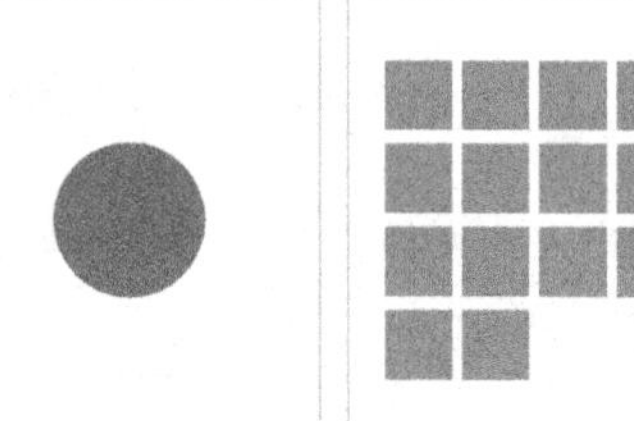

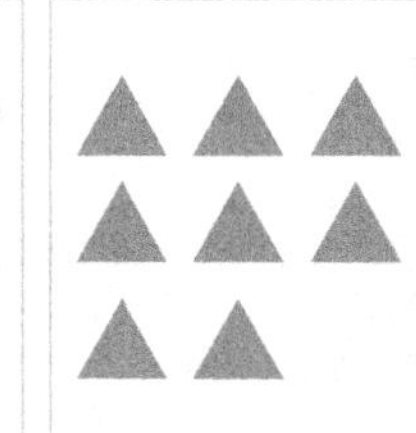

A. **B.** **C.**

K.CC.C

DAY 6
Challenge question

Draw a set that is larger than 9.

1.OA.D.8

WEEK 9

ARGOPREP.COM

VIDEO
EXPLANATIONS

2 1 3

Week 9 is all about comparing two numbers that are between 1 and 10.

You can find detailed video explanations of each problem in the book by visiting: ArgoPrep.com/ccmk

WEEK 9 : DAY 1

1. Which number is larger? 1 or 3

____

K.CC.C.7

2. Which number is larger? 9 or 4

____

K.CC.C.7

3. Which number is larger? 2 or 9

____

K.CC.C.7

4. Which number is larger? 8 or 2

____

K.CC.C

5. Which number is larger? 6 or 1

____

K.CC.C

6. Which number is larger? 3 or 5

____

K.CC.C

Represent each number with objects. Circle the larger group.

WEEK 9 : DAY 2

Which number is smaller? 2 or 4

____

K.CC.C.7

4. Which number is smaller? 10 or 3

____

K.CC.C.7

Which number is smaller? 4 or 7

____

K.CC.C.7

5. Which number is smaller? 7 or 8

____

K.CC.C.7

Which number is smaller? 9 or 6

____

K.CC.C.7

6. Which number is smaller? 5 or 6

____

K.CC.C.7

Represent each number with objects and circle the smaller group.

WEEK 9 : DAY 3

1. Which number is larger? 1 or 9

____

K.CC.C.7

2. Which number is larger? 9 or 5

____

K.CC.C.7

3. Which number is larger? 7 or 2

____

K.CC.C.7

4. Which number is larger? 2 or 8

____

K.CC.C

5. Which number is larger? 9 or 3

____

K.CC.C

6. Which number is larger? 10 or 8

____

K.CC.C

Cross out the smaller number first.

WEEK 9 : DAY 4

Which number is smaller? 6 or 4

____

K.CC.C.7

Which number is smaller? 3 or 9

____

K.CC.C.7

Which number is smaller? 8 or 1

____

K.CC.C.7

4. Which number is smaller? 4 or 7

____

K.CC.C.7

5. Which number is smaller? 8 or 6

____

K.CC.C.7

6. Which number is smaller? 3 or 6

____

K.CC.C.7

Cross out the larger number first.

WEEK 9 : DAY 5

ASSESSMENT

1. Which number is smaller? 5 or 2

____

K.CC.C.7

2. Which number is larger? 7 or 10

____

K.CC.C.7

3. Which number is larger? 5 or 4

____

K.CC.C.7

4. Which number is smaller? 6 or 1

____

K.CC.C

5. Which number is larger? 4 or 5

____

K.CC.C

6. Which number is smallest? 2 or 7 or 3

____

K.CC.C.

DAY 6
Challenge question

Provide a number that is larger than 26.

1.NBT.A.1

WEEK 10

This week focuses on representing addition and subtraction with objects, fingers, mental images or drawings.

You can find detailed video explanations of each problem in the book by visiting: ArgoPrep.com/ccmk

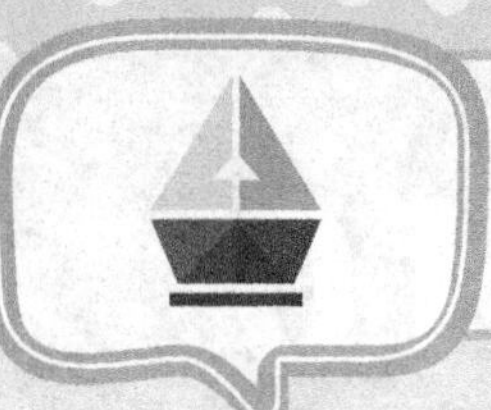

WEEK 10 : DAY 1

1. Use drawings or another method to add 1 + 9.

   A. 10
   B. 11
   C. 12
   D. 13

   K.OA.A.1

2. Use drawings or another method to add 3 + 6.

   A. 8
   B. 9
   C. 10
   D. 11

   K.OA.A.1

3. Use drawings or another method to add 2 + 4.

   A. 5
   B. 6
   C. 7
   D. 8

   K.OA.A.1

4. Use drawings or another method to ac 3 + 9.

   A. 11
   B. 12
   C. 13
   D. 14

   K.OA.A

5. Use drawings or another method to ac 7 + 3.

   A. 7
   B. 8
   C. 9
   D. 10

   K.OA.A.

6. Use drawings or another method to ad 10 + 2.

   A. 11
   B. 12
   C. 13
   D. 14

   K.OA.A.

Draw a set of objects that represent each number.

WEEK 10 : DAY 2

Use drawings or another method to add 4 + 1.

A. 5
B. 6
C. 7
D. 8

K.OA.A.1

Use drawings or another method to add 4 + 6.

A. 8
B. 9
C. 10
D. 11

K.OA.A.1

Use drawings or another method to add 5 + 8.

A. 13
B. 14
C. 15
D. 16

K.OA.A.1

4. Use drawings or another method to add 3 + 5.

A. 7
B. 8
C. 9
D. 10

K.OA.A.1

5. Use drawings or another method to add 7 + 2.

A. 6
B. 7
C. 8
D. 9

K.OA.A.1

6. Use drawings or another method to add 1 + 8.

A. 7
B. 8
C. 9
D. 10

K.OA.A.1

Count the two sets of objects that you have drawn.

WEEK 10 : DAY 3

1. Use drawings or another method to subtract 10 - 9.

K.OA.A.1

2. Use drawings or another method to subtract 7 - 3.

K.OA.A.1

3. Use drawings or another method to subtract 6 - 4.

K.OA.A.1

4. Use drawings or another method subtract 9 - 3.

K.OA.A

5. Use drawings or another method subtract 5 - 1.

K.OA.A

6. Use drawings or another method subtract 8 - 2.

K.OA.A

Draw a set of objects that represents the first number.

WEEK 10 : DAY 4

Use drawings or another method to subtract 4 - 2.

K.OA.A.1

4. Use drawings or another method to subtract 6 - 2.

K.OA.A.1

Use drawings or another method to subtract 7 - 5.

K.OA.A.1

5. Use drawings or another method to subtract 5 - 2.

K.OA.A.1

. Use drawings or another method to subtract 3 - 2.

K.OA.A.1

6. Use drawings or another method to subtract 2 - 1.

K.OA.A.1

Cross out the number of objects equal to the second number.

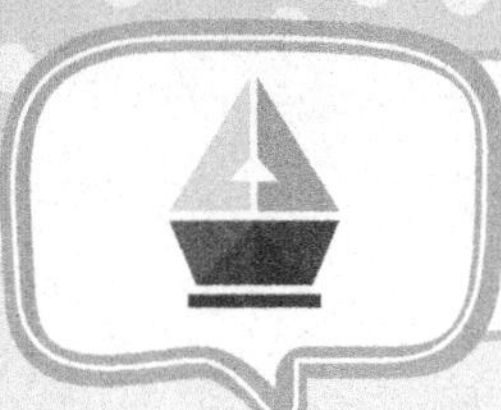

WEEK 10 : DAY 5

ASSESSMENT

1. Use drawings or another method to add 1 + 4.

K.OA.A.1

2. Use drawings or another method to add 5 + 2.

K.OA.A.1

3. Use drawings or another method to subtract 4 - 1.

K.OA.A.1

4. Use drawings or another method t subtract 9 - 2.

K.OA.A

5. Use drawings or another method to ac 8 + 6.

K.OA.A.

6. Use drawings or another method t subtract 7 - 3.

K.OA.A.

DAY 6
Challenge question

6 + 3 - 5 = ☐

1.NBT.B.2

This week you will work with addition and subtraction word problems and solve them by using drawings and objects.

You can find detailed video explanations of each problem in the book by visiting: ArgoPrep.com/ccmk

WEEK 11 : DAY 1

1. A dog has 4 legs. A bird has 2 legs. Draw a model to represent how many legs they have in all.

K.OA.A.2

2. A dog has 4 legs. A bird has 2 legs. Write a problem to represent how many legs they have in all.

____

K.OA.A.2

3. A dog has 4 legs. A bird has 2 legs. How many legs do they have in all?

____

K.OA.A.2

4. Alicia's mom bought 6 apples at the sto and her family ate 3 on the first day. Dra a model to represent how many appl were left.

K.OA.A

5. Alicia's mom bought 6 apples at the sto and her family ate 3 on the first day. Wri a problem to represent how many appl were left.

____

K.OA.A.

6. Alicia's mom bought 6 apples at the sto and her family ate 3 on the first day. Ho many apples were left?

____

K.OA.A.

Take each word problem a step at a time.

WEEK 11 : DAY 2

Our teacher had 8 pencils. Three kids need a pencil. Draw a model to represent how many she had left K.OA.A.2	4. My cousins came to visit. We had 10 kids and 4 were girls. Draw a model to show how many boys there were? K.OA.A.2
Our teacher had 8 pencils. Three kids need a pencil. Write a problem to represent how many she had left. ____________ K.OA.A.2	5. My cousins came to visit. We had 10 kids and 4 were girls. Write a problem to show how many boys there were? ____________ K.OA.A.2
Our teacher had 8 pencils. Three kids need a pencil. How many did she have left? ____________ K.OA.A.2	6. My cousins came to visit. We had 10 kids and 4 were girls. How many boys were there? ____________ K.OA.A.2

First, represent the problem with a drawing.

WEEK 11 : DAY 3

1. For my birthday, we had pizza. I ate 3 pieces of pizza and my dad ate 4. Draw a model to show how many pieces of pizza we ate together.

K.OA.A.2

4. There are six chickens and 2 cows c my grandparent's farm. Draw a model t represent how many animals there a on the farm.

K.OA.A

2. For my birthday, we had pizza. I ate 3 pieces of pizza and my dad ate 4. Write a problem to show how many pieces of pizza we ate together.

____

K.OA.A.2

5. There are six chickens and 2 cows on m grandparent's farm. Write a problem t represent how many animals there a on the farm.

____

K.OA.A.

3. For my birthday, we had pizza. I ate 3 pieces of pizza and my dad ate 4. How many pieces of pizza did we eat in all.

____

K.OA.A.2

6. There are six chickens and 2 cows on m grandparent's farm. How many anima are there on the farm?

____

K.OA.A.

TIP of the DAY

Second, add more pictures or cross out some listed.

WEEK 11 : DAY 4

My sister has 3 stuffed bears. I have 6 stuffed bears. Draw a model to represent how many bears we have in all.

K.OA.A.2

4. Last summer, my brother read 4 books. My sister read 8 books. Draw a model to see who read more books.

K.OA.A.2

My sister has 3 stuffed bears. I have 6 stuffed bears. Write a problem to represent how many bears we have in all.

____

K.OA.A.2

5. Last summer, my brother read 4 books. My sister read 8 books. Who read more books?

____

K.OA.A.2

My sister has 3 stuffed bears. I have 6 stuffed bears. How many bears do we have in all?

____

K.OA.A.2

6. Last summer, my brother read 4 books. My sister read 8 books. How many more books did my sister read?

____

K.OA.A.2

Third, count the objects left.

WEEK 11 : DAY 5

ASSESSMENT

1. There are six cars in your toybox and four trucks. Draw a model to represent how many vehicles there are in all.

K.OA.A.2

4. Write your own problem that would b solved using addition!

____

K.OA.A

2. There are six cars in your toybox and four trucks. Write a problem to represent how many vehicles there are in all.

____

K.OA.A.2

5. Write your own problem that would b solved using addition!

____

K.OA.A.

3. There are six cars in your toybox and four trucks. How many vehicles are there are in all?

____

K.OA.A.2

6. Write your own problem that would b solved using subtraction!

____

K.OA.A.

DAY 6 Challenge question

Melissa got a new book for her birthday. One day, she read 6 pages. On the next day, she read 4 pages. Finally, she read 7 pages. If the book has 19 pages, has she finished the book?

1.NBT.B.3

In week 12, we will continue to use diagrams to help us solve addition and subtraction word problems.

You can find detailed video explanations of each problem in the book by visiting: ArgoPrep.com/ccmk

WEEK 12 : DAY 1

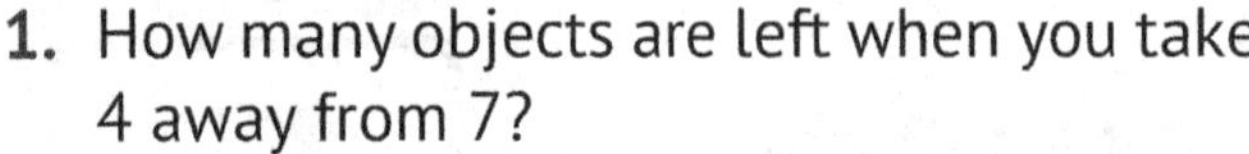

1. How many objects are left when you take 4 away from 7?

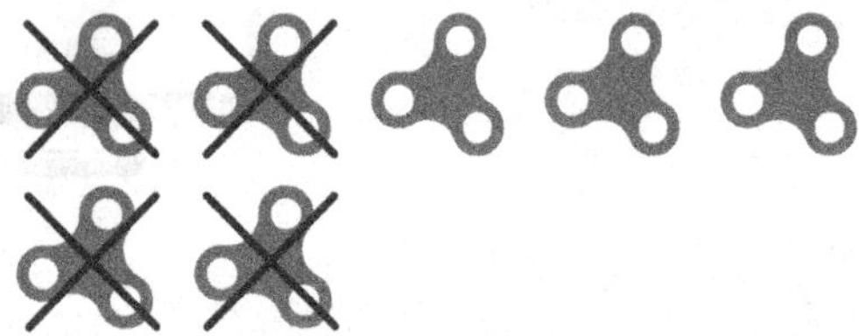

______________ K.OA.A.3

2. How many objects are left when you take 8 away from 9?

______________ K.OA.A.3

3. How many objects are left when you take 3 away from 4?

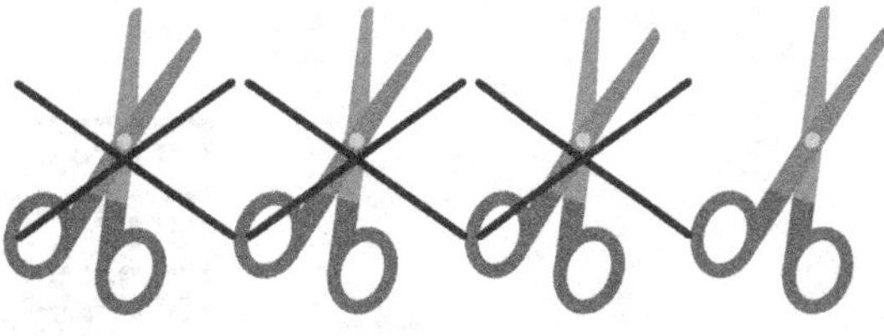

______________ K.OA.A.3

4. How many objects are left when you tak
5 away from 6?

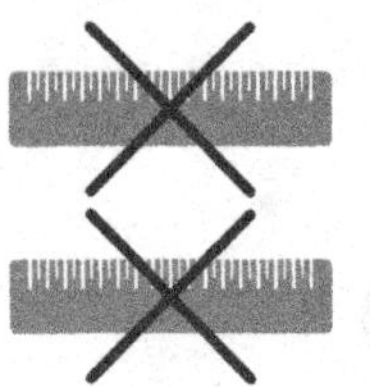
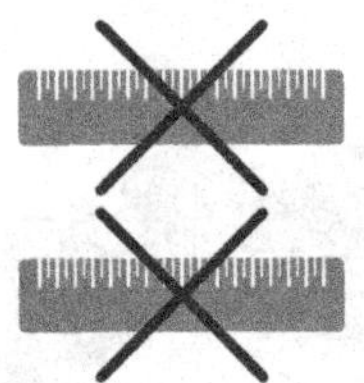

______________ K.OA.A

5. How many objects are left when you tak
1 away from 2?

______________ K.OA.A.

6. How many objects are left when you tak
6 away from 12?

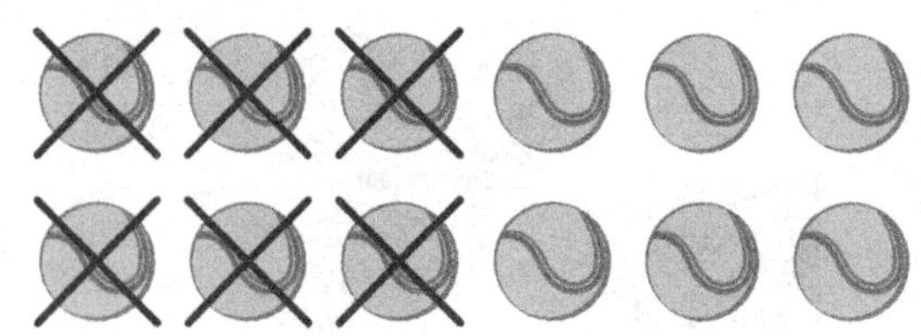

______________ K.OA.A.

TIP of the DAY

Count the objects that do not have an X drawn through them.

WEEK 12 : DAY 2

How many objects do you need to add to the drawing to get 7?

____________________ K.OA.A.3

How many objects do you need to add to the drawing to get 5?

____________________ K.OA.A.3

How many objects do you need to add to the drawing to get 8?

____________________ K.OA.A.3

4. How many objects do you need to add to the drawing to get 9?

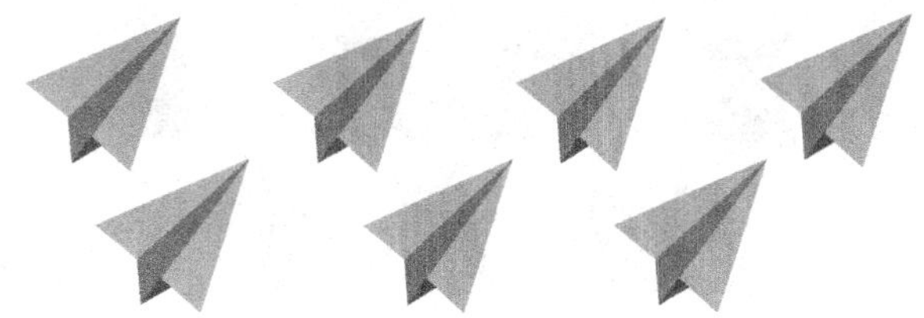

____________________ K.OA.A.3

5. How many objects do you need to add to the drawing to get 6?

____________________ K.OA.A.3

6. How many objects do you need to add to the drawing to get 8?

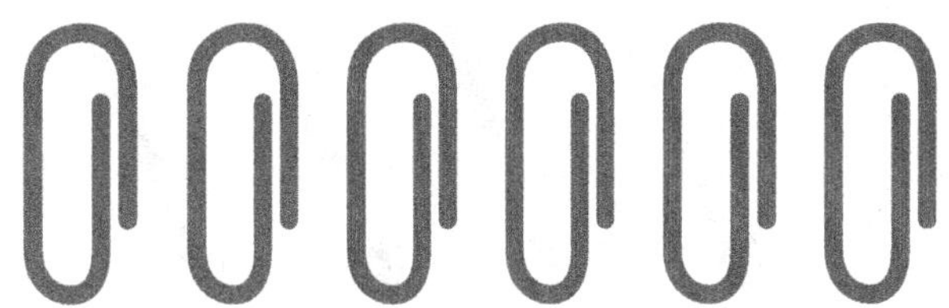

____________________ K.OA.A.3

Draw objects to get to the number listed. Count the number of objects you need to draw.

WEEK 12 : DAY 3

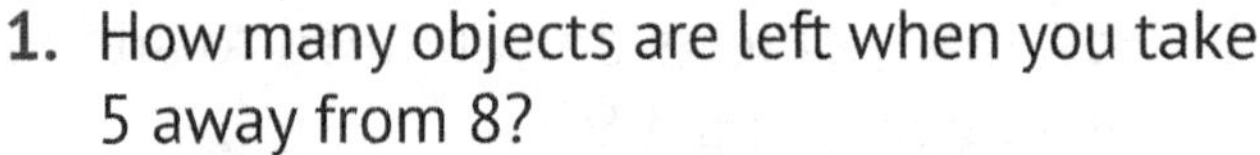

1. How many objects are left when you take 5 away from 8?

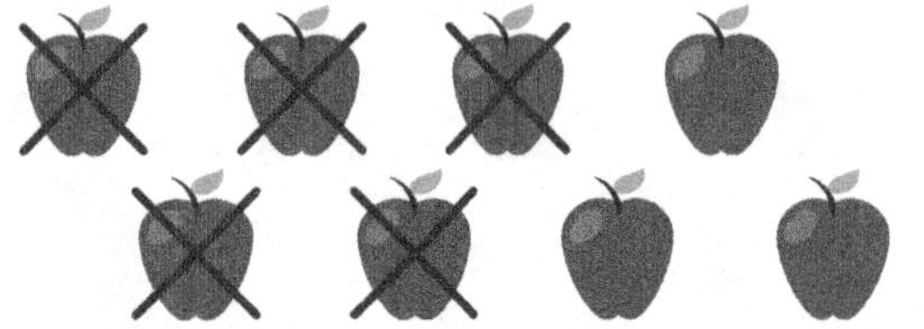

______________________ K.OA.A.3

2. How many objects are left when you take 2 away from 7?

______________________ K.OA.A.3

3. How many objects do you need to add to the drawing to get 8?

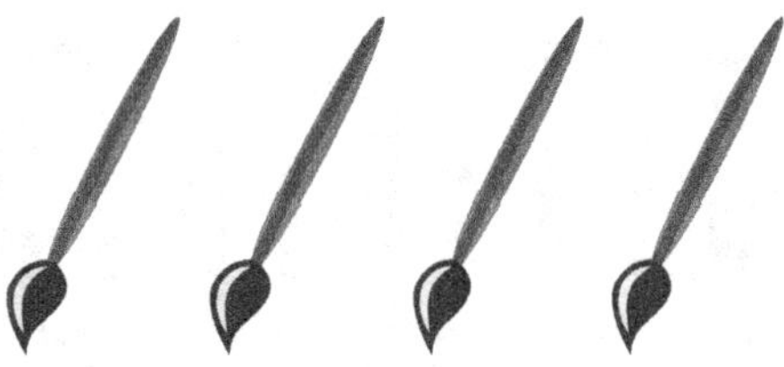

______________________ K.OA.A.3

4. How many objects are left when you tak
3 away from 9?

______________________ K.OA.A

5. How many objects do you need to add t
the drawing to get 9?

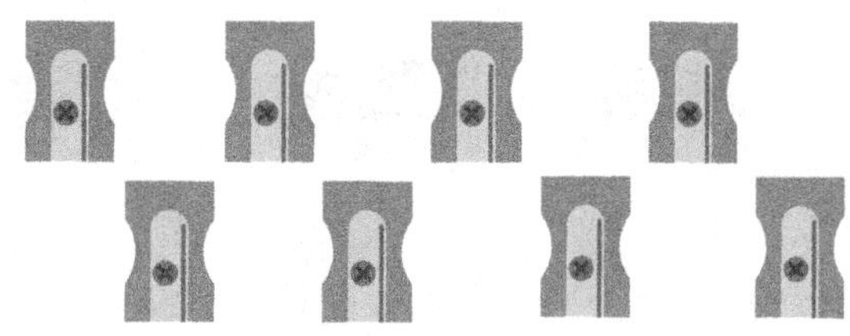

______________________ K.OA.A

6. How many objects do you need to add t
the drawing to get 7?

______________________ K.OA.A.

When adding and subtracting are mixed, pay careful attention to the words of the problem.

WEEK 12 : DAY 4

. What number do you need to combine with 1 to get 3?

A. 1
B. 2
C. 3
D. 4

K.OA.A.3

4. What number do you need to combine with 3 to get 4?

A. 1
B. 2
C. 3
D. 4

K.OA.A.3

. What number do you need to combine with 4 to get 6?

A. 1
B. 2
C. 3
D. 4

K.OA.A.3

5. What number do you need to combine with 3 to get 7?

A. 1
B. 2
C. 3
D. 4

K.OA.A.3

. What number do you need to combine with 2 to get 8?

A. 5
B. 6
C. 7
D. 8

K.OA.A.3

6. What number do you need to combine with 2 to get 5?

A. 1
B. 2
C. 3
D. 4

K.OA.A.3

TIP of the DAY

Take away means subtraction; add to means addition.

WEEK 12 : DAY 5

ASSESSMENT

1. What number do you need to combine with 4 to get 9?

   A. 5
   B. 6
   C. 7
   D. 8

   K.OA.A.3

2. What number do you need to combine with 1 to get 6?

   A. 2
   B. 3
   C. 4
   D. 5

   K.OA.A.3

3. What number do you need to combine with 6 to get 8?

   A. 1
   B. 2
   C. 3
   D. 4

   K.OA.A.3

4. What number do you need to combir with 5 to get 7?

   A. 2
   B. 3
   C. 4
   D. 5

   K.OA.A

5. Draw 4 objects. Circle the two sets wit the same number of objects.

   K.OA.A.

6. What number can be doubled to get 6?

   A. 1
   B. 2
   C. 3
   D. 4

   K.OA.A.

DAY 6 Challenge question

If you start with 1 object and you get 7 more objects, how many more objects do you need to get to 16?

1.NBT.C.4

Week 13 focuses on working with numbers from 1 to 9, and finding the number that makes 10 when added to the given number.

You can find detailed video explanations of each problem in the book by visiting: ArgoPrep.com/ccmk

WEEK 13 : DAY 1

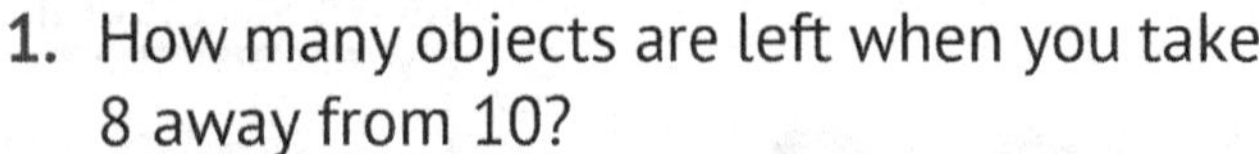

1. How many objects are left when you take 8 away from 10?

______________________ K.OA.A.4

2. How many objects are left when you take 2 away from 10?

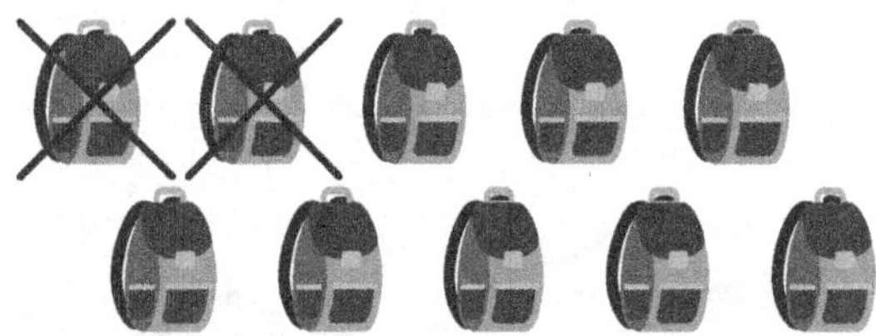

______________________ K.OA.A.4

3. How many objects are left when you take 3 away from 10?

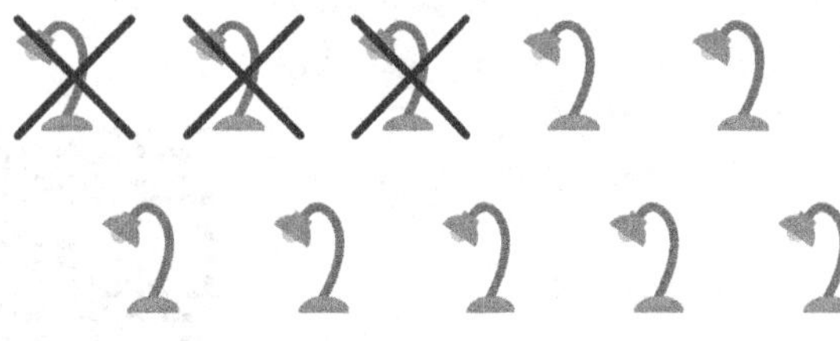

______________________ K.OA.A.4

4. How many objects are left when you tak
5 away from 10?

______________________ K.OA.A

5. How many objects are left when you tak
4 away from 10?

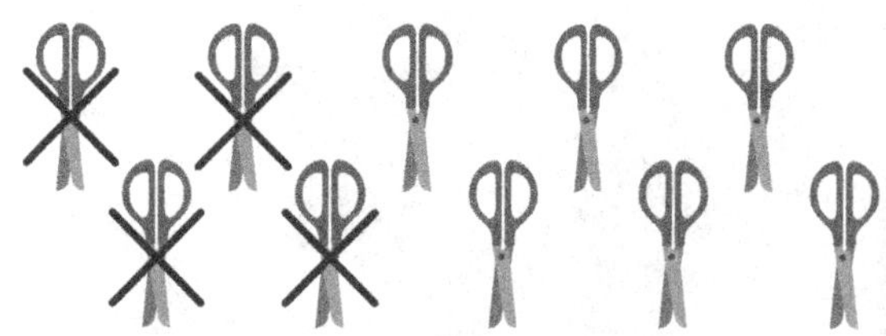

______________________ K.OA.A

6. How many objects are left when you tak
1 away from 10?

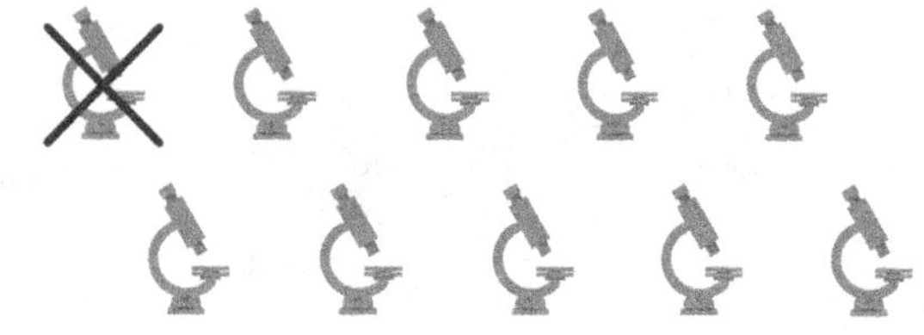

______________________ K.OA.A.

It can be helpful to memorize the combinations of numbers that add up to 10.

WEEK 13 : DAY 2

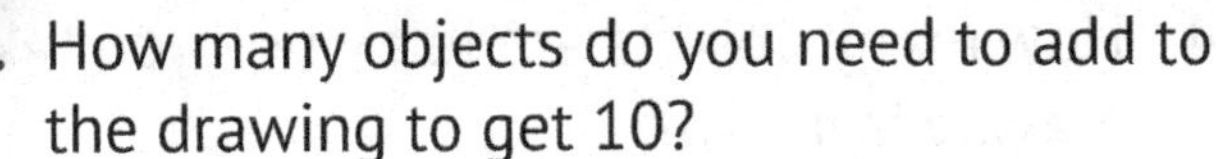

How many objects do you need to add to the drawing to get 10?

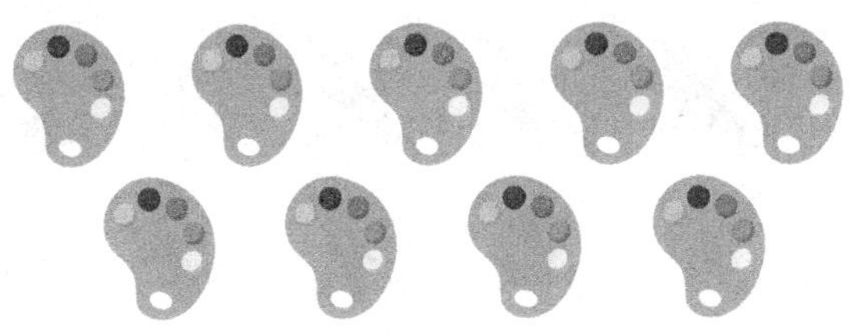

______________________ K.OA.A.4

How many objects do you need to add to the drawing to get 10?

______________________ K.OA.A.4

How many objects do you need to add to the drawing to get 10?

______________________ K.OA.A.4

4. How many objects do you need to add to the drawing to get 10?

______________________ K.OA.A.4

5. How many objects do you need to add to the drawing to get 10?

______________________ K.OA.A.4

6. How many objects do you need to add to the drawing to get 10?

______________________ K.OA.A.4

TIP of the DAY

1 + 9 = ☐ *3 + 7 =* ☐ *5 + 5 =* ☐

2 + 8 = ☐ *4 + 6 =* ☐

WEEK 13 : DAY 3

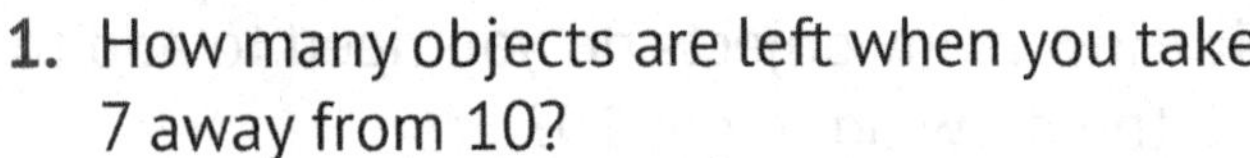

1. How many objects are left when you take 7 away from 10?

______________ K.OA.A.4

2. How many objects are left when you take 9 away from 10?

______________ K.OA.A.4

3. How many objects do you need to add to the drawing to get 10?

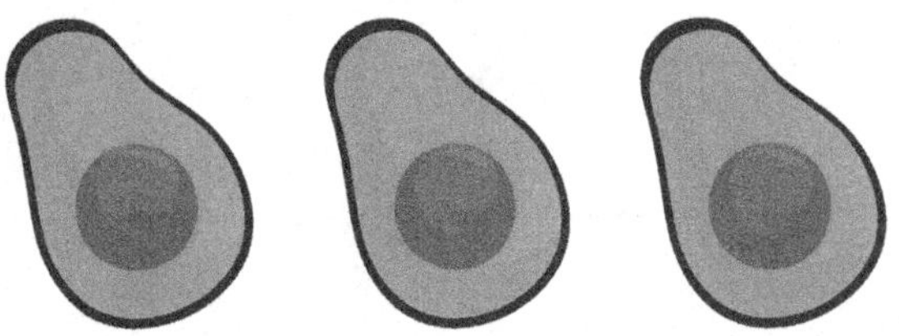

______________ K.OA.A.4

4. How many objects are left when you tak 6 away from 10?

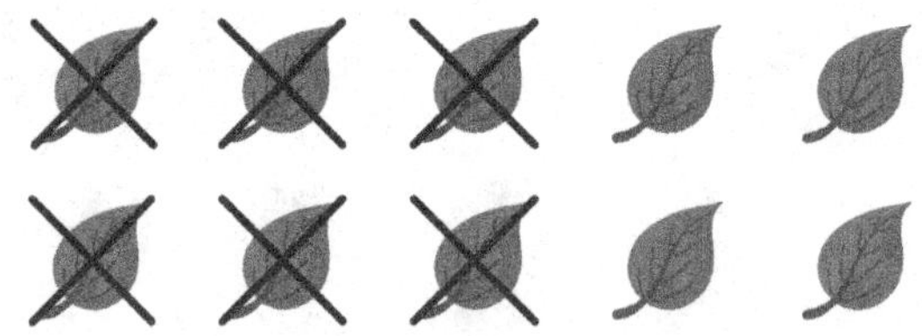

______________ K.OA.A.

5. How many objects do you need to add t the drawing to get 10?

______________ K.OA.A.

6. How many objects do you need to add t the drawing to get 10?

______________ K.OA.A.

You can rewrite these problems as addition and subtraction problems if you wish.

WEEK 13 : DAY 4

What number do you need to combine with 3 to get 10?

A. 5
B. 6
C. 7
D. 8

K.OA.A.4

What number do you need to combine with 5 to get 10?

A. 3
B. 4
C. 5
D. 6

K.OA.A.4

What number do you need to combine with 2 to get 10?

A. 6
B. 7
C. 8
D. 9

K.OA.A.4

4. What number do you need to combine with 7 to get 10?

A. 1
B. 2
C. 3
D. 4

K.OA.A.4

5. What number do you need to combine with 9 to get 10?

A. 1
B. 2
C. 3
D. 4

K.OA.A.4

6. What number do you need to combine with 4 to get 10?

A. 4
B. 5
C. 6
D. 7

K.OA.A.4

TIP *of the* DAY

Consider your number combinations to get to 10.

WEEK 13 : DAY 5

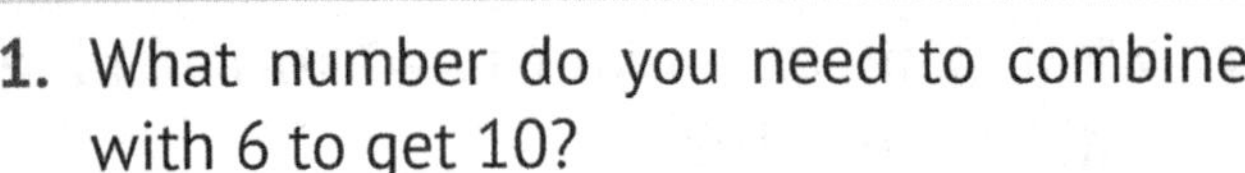

ASSESSMENT

1. What number do you need to combine with 6 to get 10?

A. 3
B. 4
C. 5
D. 6

K.OA.A.4

2. What number do you need to combine with 1 to get 10?

A. 6
B. 7
C. 8
D. 9

K.OA.A.4

3. What number do you need to combine with 10 to get 10?

A. 0
B. 1
C. 2
D. 3

K.OA.A.4

4. What number do you need to combin with 8 to get 10?

A. 1
B. 2
C. 3
D. 4

K.OA.A

5. Circle the two sets with the same numbe of objects.

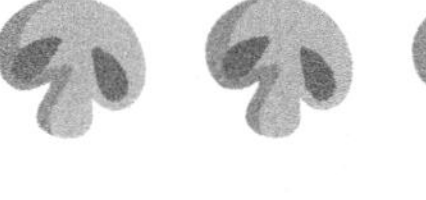
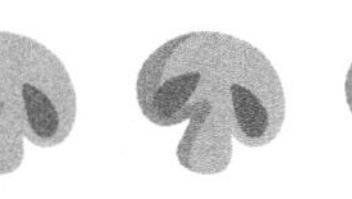
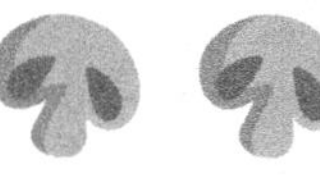
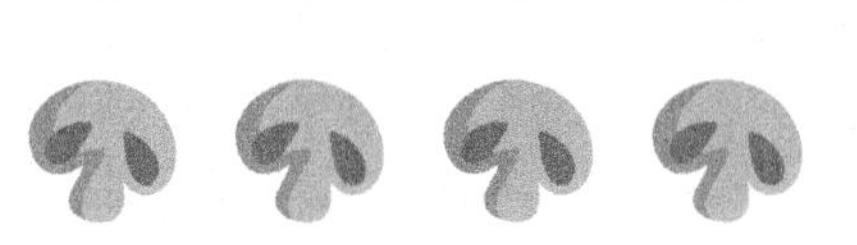

K.OA.A.

6. What number can be doubled to get 20

____

K.OA.A.

DAY 6
Challenge question

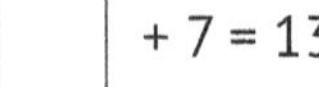
☐ + 7 = 13

1.NBT.C.5

Ready for week 14? This week is all about getting comfortable adding and subtracting within 5.

You can find detailed video explanations of each problem in the book by visiting:
ArgoPrep.com/ccmk

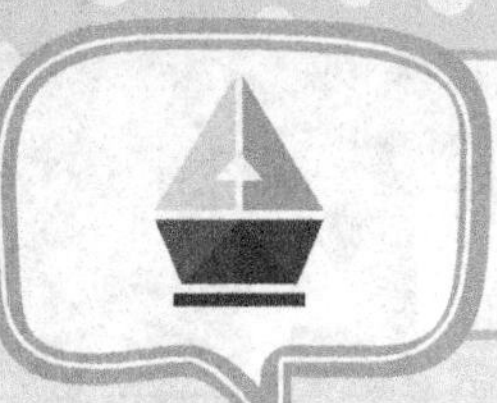

WEEK 14 : DAY 1

Fluently add and subtract within 5

1. 2 + 1 = ☐

K.OA.A.5

2. 5 + 2 = ☐

K.OA.A.5

3. 1 + 2 = ☐

K.OA.A.5

4. 4 + 3 = ☐

K.OA.A

5. 3 + 5 = ☐

K.OA.A.

6. 2 + 4 = ☐

K.OA.A.

You may use your fingers to help you calculate the answers to these problems.

WEEK 14 : DAY 2

. 6 + 2 = ☐

K.OA.A.5

4. 8 + 3 = ☐

K.OA.A.5

. 7 + 2 = ☐

K.OA.A.5

5. 4 + 1 = ☐

K.OA.A.5

. 3 + 3 = ☐

K.OA.A.5

6. 2 + 2 = ☐

K.OA.A.5

It may help to use a number line to add or subtract numbers.

WEEK 14 : DAY 3

1. 5 + 1 = ☐

K.OA.A.5

2. 8 + 4 = ☐

K.OA.A.5

3. 3 + 4 = ☐

K.OA.A.5

4. 6 + 3 = ☐

K.OA.A

5. 2 + 4 = ☐

K.OA.A

6. 4 + 2 = ☐

K.OA.A.

Count to 10: 1 2 3 4 5 6 7 8 9 10

WEEK 14 : DAY 4

1. 1 + 4 = ?

A. 4
B. 5
C. 6
D. 7

K.OA.A.5

2. 2 + 3 = ?

A. 5
B. 6
C. 7
D. 8

K.OA.A.5

3. 9 + 2 = ?

A. 9
B. 10
C. 11
D. 12

K.OA.A.5

4. 4 + 5 = ?

A. 6
B. 7
C. 8
D. 9

K.OA.A.5

5. 7 + 1 = ?

A. 7
B. 8
C. 9
D. 10

K.OA.A.5

6. 3 + 2 = ?

A. 3
B. 4
C. 5
D. 6

K.OA.A.5

TIP of the DAY

If you need to take a break when working, stand up and get a drink of water.

WEEK 14 : DAY 5

ASSESSMENT

1. 4 + 4 = ?

A. 6
B. 7
C. 8
D. 9

K.OA.A.5

2. 8 + 5 = ?

A. 12
B. 13
C. 14
D. 15

K.OA.A.5

3. 2 + 5 = ?

A. 7
B. 8
C. 9
D. 10

K.OA.A.5

4. 1 + 3 = ?

A. 2
B. 3
C. 4
D. 5

K.OA.A

5. 5 + 4 = ?

A. 7
B. 8
C. 9
D. 10

K.OA.A.

6. 9 + 3 = ?

A. 9
B. 10
C. 11
D. 12

K.OA.A.

DAY 6

Challenge question

5 + 1 + ☐ = 14

1.NBT.C.6

Week 15 works on composing and decomposing numbers from 11 to 19 into ten ones and some further ones.

You can find detailed video explanations of each problem in the book by visiting: ArgoPrep.com/ccmk

WEEK 15 : DAY 1

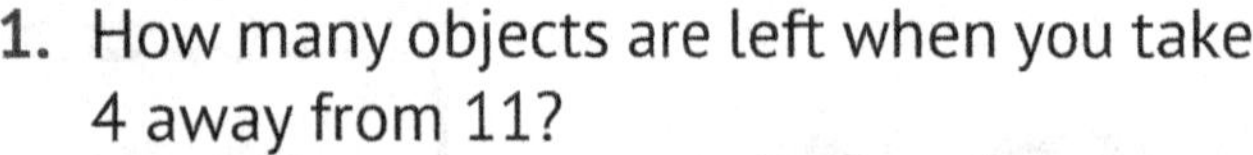

1. How many objects are left when you take 4 away from 11?

______________________ K.NBT.A.1

2. How many objects are left when you take 2 away from 16?

______________________ K.NBT.A.1

3. How many objects are left when you take 5 away from 18?

______________________ K.NBT.A.1

4. How many objects are left when you tak
8 away from 12?

______________________ K.NBT.A

5. How many objects are left when you tak
9 away from 17?

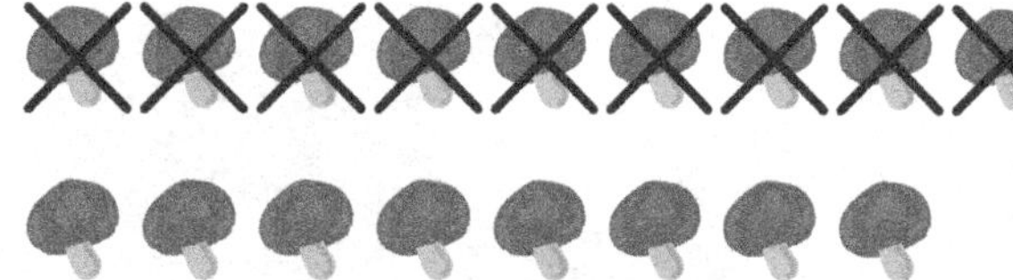

______________________ K.NBT.A

6. How many objects are left when you tak
1 away from 13?

______________________ K.NBT.A

Count the objects left in the drawing.

WEEK 15 : DAY 2

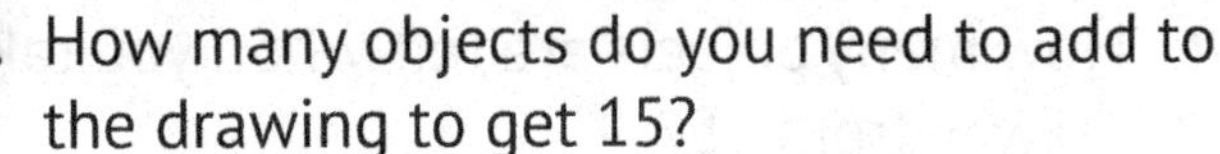

How many objects do you need to add to the drawing to get 15?

____________________ K.NBT.A.1

How many objects do you need to add to the drawing to get 17?

____________________ K.NBT.A.1

How many objects do you need to add to the drawing to get 11?

____________________ K.NBT.A.1

4. How many objects do you need to add to the drawing to get 16?

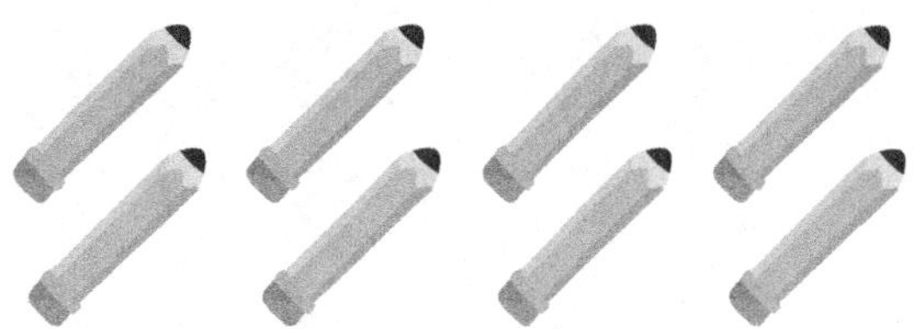

____________________ K.NBT.A.1

5. How many objects do you need to add to the drawing to get 12?

____________________ K.NBT.A.1

6. How many objects do you need to add to the drawing to get 14?

____________________ K.NBT.A.1

It may help to draw the total number and cross out the objects you already have.

WEEK 15 : DAY 3

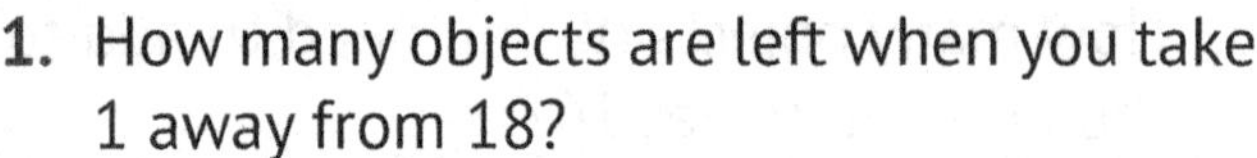

1. How many objects are left when you take 1 away from 18?

_______________ K.NBT.A.1

2. How many objects are left when you take 6 away from 14?

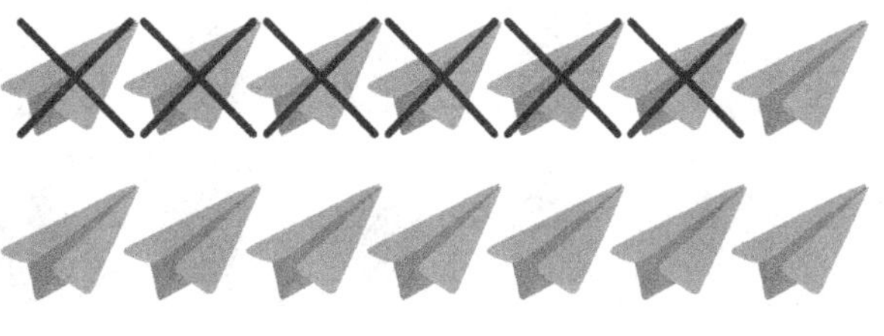

_______________ K.NBT.A.1

3. How many objects do you need to add to the drawing to get 18?

_______________ K.NBT.A.1

4. How many objects are left when you tak 7 away from 15?

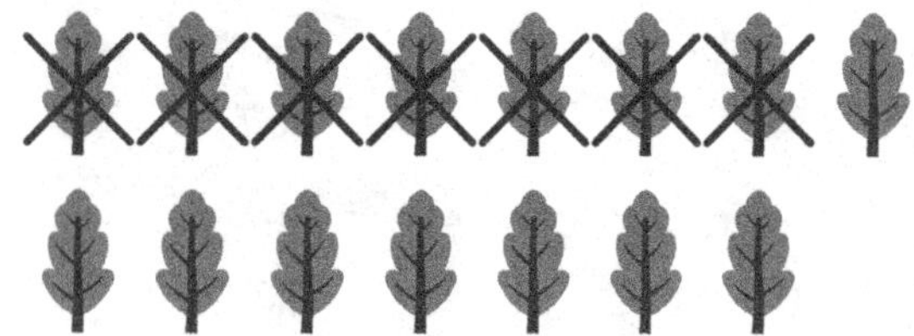

_______________ K.NBT.A.

5. How many objects do you need to add t the drawing to get 19?

_______________ K.NBT.A.

6. How many objects do you need to add t the drawing to get 13?

_______________ K.NBT.A.

Let's count to 20 to review!
1 2 3 4 5 6 7 8 9 10 11 12 13 14 15 16 17 18 19 20

WEEK 15 : DAY 4

1. What number do you need to combine with 11 to get 13?

____

K.NBT.A.1

4. What number do you need to combine with 17 to get 18?

____

K.NBT.A.1

2. What number do you need to combine with 15 to get 17?

____

K.NBT.A.1

5. What number do you need to combine with 9 to get 11?

____

K.NBT.A.1

3. What number do you need to combine with 12 to get 15?

____

K.NBT.A.1

6. What number do you need to combine with 13 to get 16?

____

K.NBT.A.1

Think about these problems as an addition or subtraction problem.

WEEK 15 : DAY 5

ASSESSMENT

1. What number do you need to combine with 12 to get 16?

____

K.NBT.A.1

2. What number do you need to combine with 15 to get 19?

____

K.NBT.A.1

3. What number do you need to combine with 10 to get 12?

____

K.NBT.A.1

4. What number do you need to combin with 8 to get 11?

____

K.NBT.A

5. What number do you need to combin with 9 to get 14?

____

K.NBT.A.

6. What number do you need to combin with 2 to get 20?

____

K.NBT.A.

DAY 6 *Challenge question*

What three numbers can you combine to get 15?

1.MD.A.1

Time to move on to measurement! This week is all about knowing when to use length or weight.

You can find detailed video explanations of each problem in the book by visiting:
ArgoPrep.com/ccmk

WEEK 16 : DAY 1

Circle length or weight.

1. What would you use to measure how long a shoe is?

   A. Length

   B. Weight

   K.MD.A.1

2. What would you use to measure how heavy an apple is?

   A. Length

   B. Weight

   K.MD.A.1

3. What would you use to measure how heavy a book is?

   A. Length

   B. Weight

   K.MD.A.1

4. What would you use to measure ho long a bed is?

   A. Length

   B. Weight

   K.MD.A.

5. What would you use to measure how ta a child is?

   A. Length

   B. Weight

   K.MD.A.

6. What would you use to measure ho heavy a box is?

   A. Length

   B. Weight

   K.MD.A.

Remember to measure length you use a ruler..

WEEK 16 : DAY 2

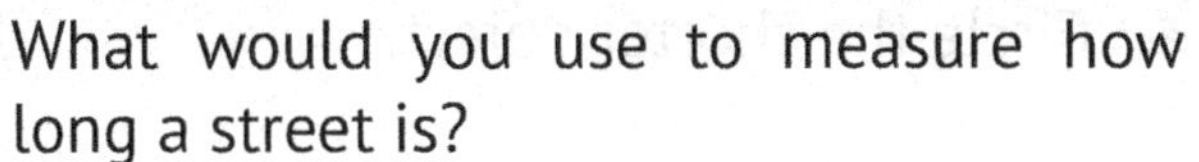

What would you use to measure how long a street is?

A. Length

B. Weight

K.MD.A.1

What would you use to measure how heavy a chair is?

A. Length

B. Weight

K.MD.A.1

What would you use to measure how tall a tree is?

A. Length

B. Weight

K.MD.A.1

4. What would you use to measure how heavy a plant is?

A. Length

B. Weight

K.MD.A.1

5. What would you use to measure how long a chalkboard is?

A. Length

B. Weight

K.MD.A.1

6. What would you use to measure how tall a fireplace is?

A. Length

B. Weight

K.MD.A.1

To measure weight you use a scale.

WEEK 16 : DAY 3

1. Circle the shorter object.

   A. A bookshelf

   B. A traffic light

   K.MD.A.2

2. Circle the shorter object.

   A. A cake

   B. A cereal box

   K.MD.A.2

3. Circle the shorter object.

   A. A cat

   B. A giraffe

   K.MD.A.2

4. Circle the shorter object.

   A. A snowman

   B. A mountain

   K.MD.A

5. Circle the shorter object.

   A. A squirrel

   B. A mouse

   K.MD.A.

6. Circle the shorter object.

   A. A dump truck

   B. A passenger car

   K.MD.A.

Think about which object would be easier to lift.

WEEK 16 : DAY 4

. Circle the heavier object.

A. An elephant

B. A caterpillar

K.MD.A.2

. Circle the heavier object.

A. A chair

B. A sofa

K.MD.A.2

. Circle the heavier object.

A. A leaf

B. A carrot

K.MD.A.2

4. Circle the heavier object.

A. A piece of paper

B. A laptop

K.MD.A.2

5. Circle the heavier object.

A. A pair of pants

B. A sock

K.MD.A.2

6. Circle the heavier object.

A. A cow

B. A container of orange juice

K.MD.A.2

Think about which object would be harder to lift.

WEEK 16 : DAY 5

ASSESSMENT

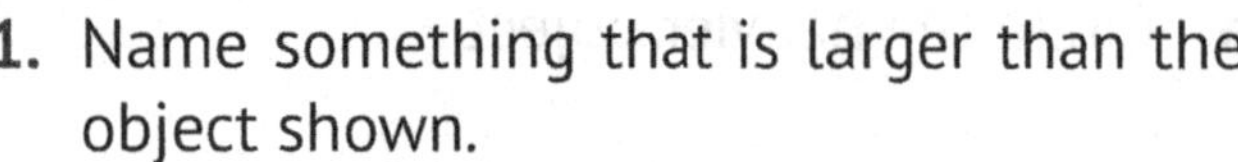

1. Name something that is larger than the object shown.

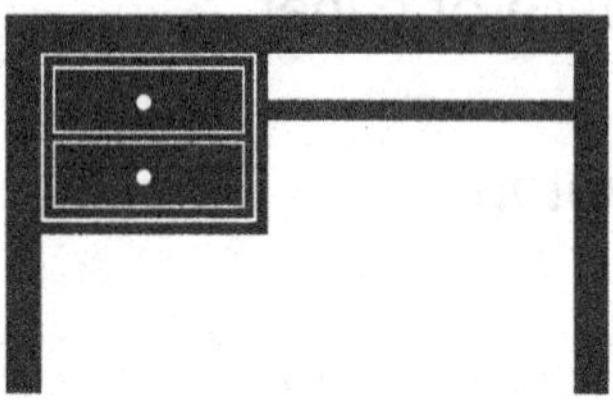

____

K.MD.A.2

4. Name something that is larger than th object shown.

____

K.MD.A.

2. Name something that is larger than the object shown.

____

K.MD.A.2

5. Name something that is larger than th object shown.

____

K.MD.A.

3. Name something that is larger than the object shown.

____

K.MD.A.2

6. Name something that is larger than th object shown.

____

K.MD.A.

DAY 6
Challenge question

What object would take longer to travel down: a street or a house?

1.MD.A.2

Week 17 is all about classifying objects into given categories

You can find detailed video explanations of each problem in the book by visiting: ArgoPrep.com/ccmk

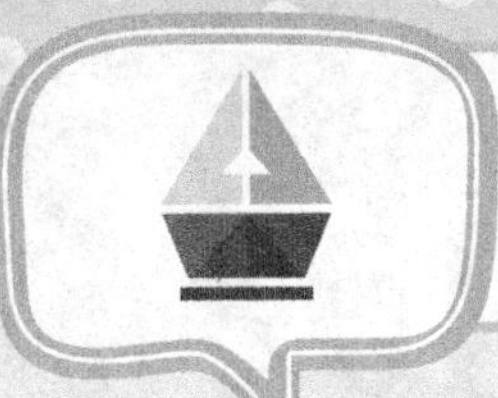

WEEK 17 : DAY 1

Classify objects into given categories; count the numbers of objects in each category and sort the categories by count.
Use the data set below to answer the following questions.

1. Which object has the least amount?

A. **B.** **C.** **D.**

K.MD.B.3

2. How many balls are there?

A. 4
B. 5
C. 6
D. 7

K.MD.B.3

3. How many dolls are there?

A. 2
B. 3
C. 4
D. 5

K.MD.B.

4. How many more jump ropes than dol are there?

A. 1
B. 2
C. 3
D. 4

K.MD.B.

5. Which two objects have the most amoun

A. **B.** **C.** **D.**

K.MD.B.

6. How many objects are there in all?

A. 18
B. 19
C. 20
D. 21

K.MD.B.

TIP of the DAY

Count each object and write the number under the object.

WEEK 17 : DAY 2

se the data set below to answer the ollowing questions.

. Which animal has the most?

A. B. C. D.

K.MD.B.3

. Which animal has the least?

A. B. C. D.

K.MD.B.3

3. Which animals have 4 legs?

A. B. C. D.

K.MD.B.3

4. Which animals have no legs?

A. B. C. D.

K.MD.B.3

5. How many snakes are there?

A. 2
B. 3
C. 4
D. 5

K.MD.B.3

6. How many turtles and frogs are there?

A. 3
B. 4
C. 6
D. 7

K.MD.B.3

Think about what you know about each type of animal.

WEEK 17 : DAY 3

Use the data set below to answer the following questions.

1. How many apples are there?

A. 4
B. 5
C. 6
D. 7

K.MD.B.3

2. How many fruit are there in all?

A. 21
B. 23
C. 25
D. 27

K.MD.B.3

3. Which fruit is the most?

 **A.**
 **B.**
 **C.**
 **D.**

K.MD.B.

4. Which fruit has the least?

 **A.**
 **B.**
 **C.**
 **D.**

K.MD.B.

5. How many oranges and apples are there

A. 8
B. 9
C. 10
D. 11

K.MD.B.

6. How many more apples are there tha oranges?

A. 1
B. 2
C. 3
D. 4

K.MD.B.

TIP of the DAY

Use numbers instead of shapes to calculate the answers.

WEEK 17 : DAY 4

se the data set below to answer the llowing questions.

3. Which color has the most?

A. B. C. D. K.MD.B.3

4. Which color has 4 items?

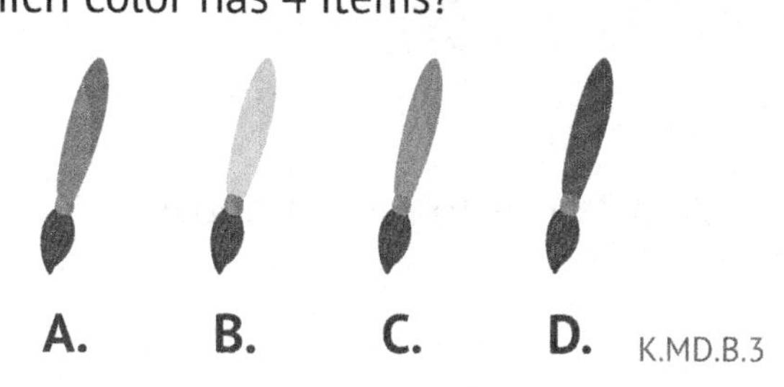

A. B. C. D. K.MD.B.3

. How many paintbrushes are there in all?

A. 11
B. 12
C. 13
D. 14

K.MD.B.3

5. **How many purple paintbrushes are there?**

A. 1
B. 2
C. 3
D. **4**

K.MD.B.3

. Which two colors are the same amount?

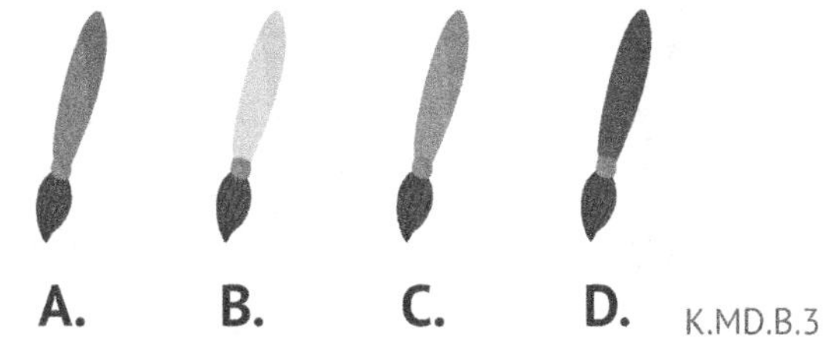

A. B. C. D. K.MD.B.3

6. **Which two colors have the least amount?**

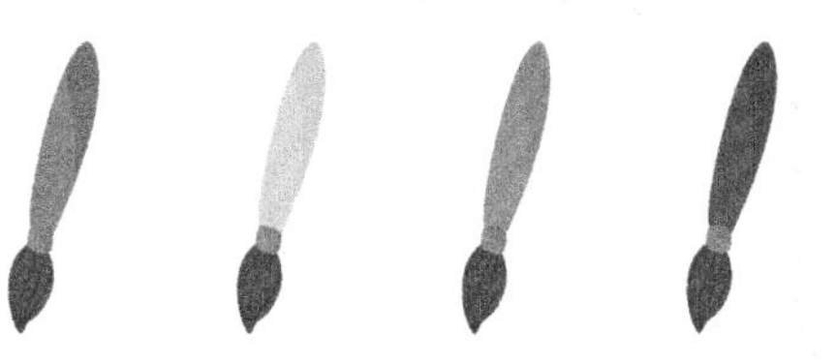

A. B. C. D. K.MD.B.3

The size of the object does not impact how many of those there are.

WEEK 17 : DAY 5

ASSESSMENT

Use the data set below to answer the following questions.

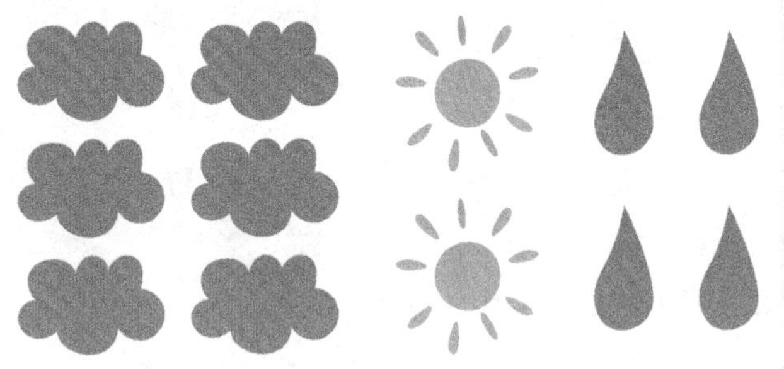

1. Add 3 suns to the set above.

K.MD.B.3

2. How many suns are there now?

A. 2
B. 4
C. 5
D. 6

K.MD.B.3

3. Which item has the most items?

A. B. C.

K.MD.B.

4. Which item has the least?

A. B. C.

K.MD.B.

5. Which item has 4 items?

A. B. C.

K.MD.B.

6. How many items are there in all?

A. 15
B. 13
C. 12
D. 10

K.MD.B.

DAY 6
Challenge question

What time is it on this clock?

____

1.MD.B.3

WEEK 18

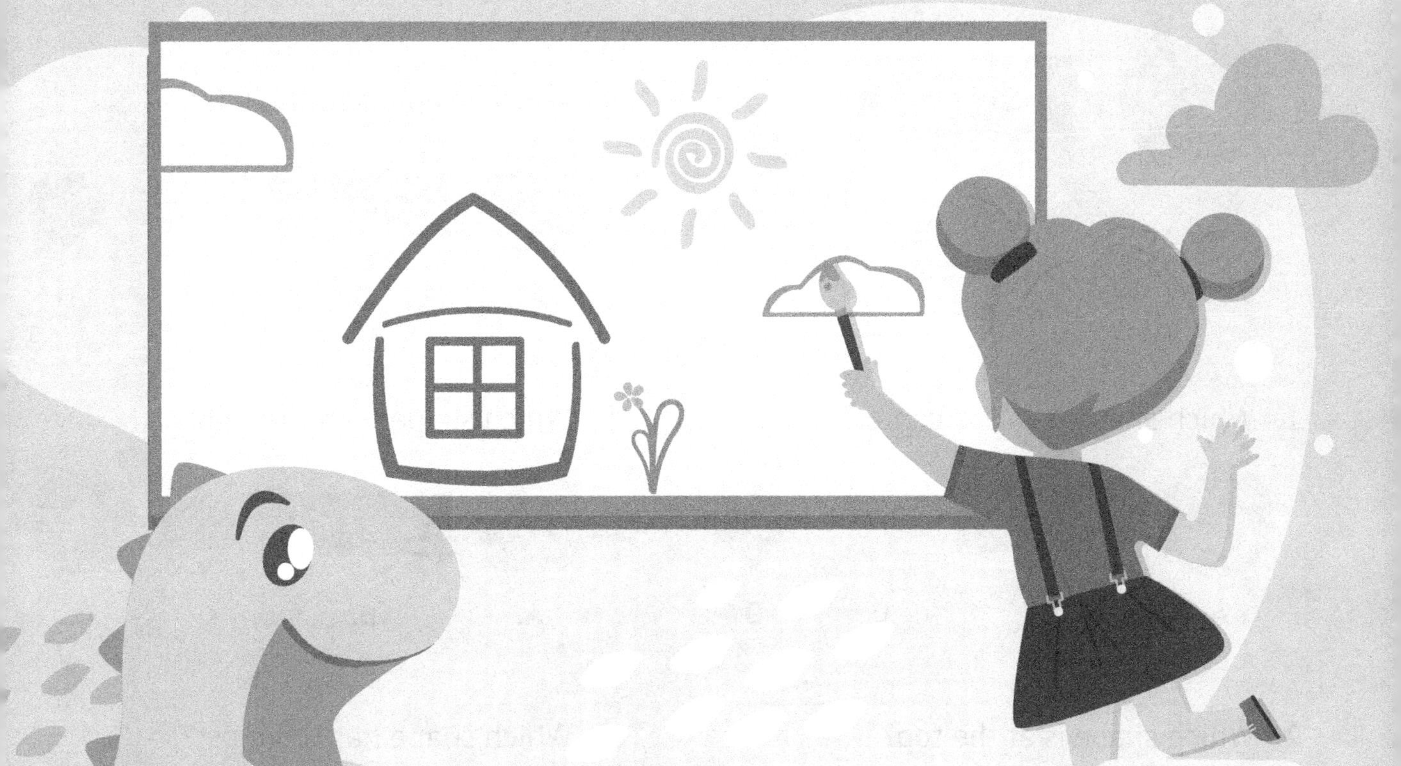

In week 18 we will be identifying and describing various shapes.

You can find detailed video explanations of each problem in the book by visiting: ArgoPrep.com/ccmk

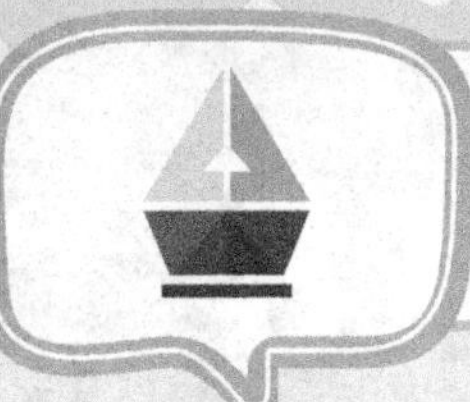

WEEK 18 : DAY 1

Use the data set below to answer the following questions.

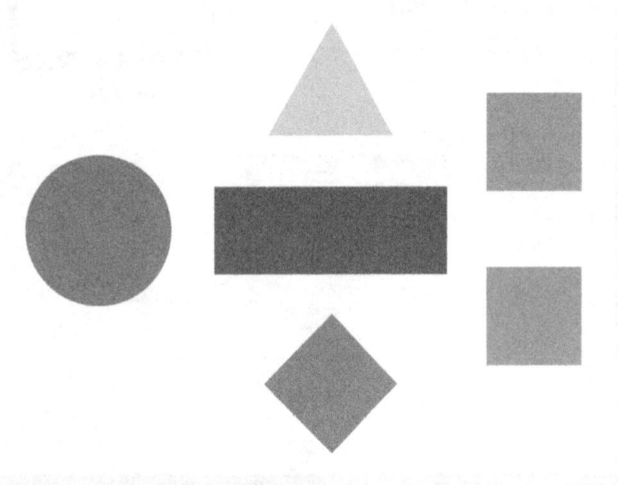

1. Which shape is in the center?

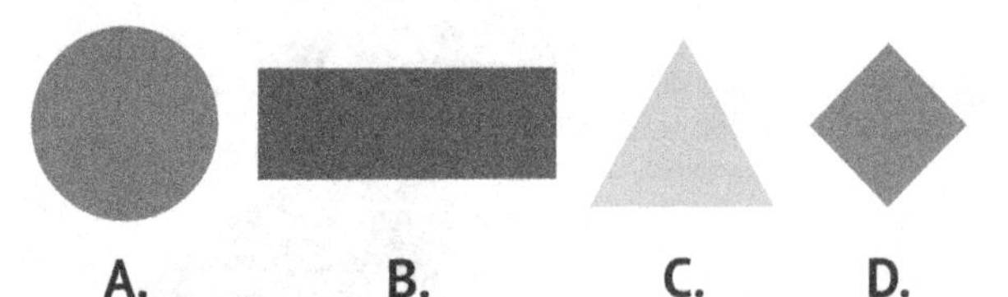

A. **B.** **C.** **D.**

K.G.A.1

2. Which shape is at the top?

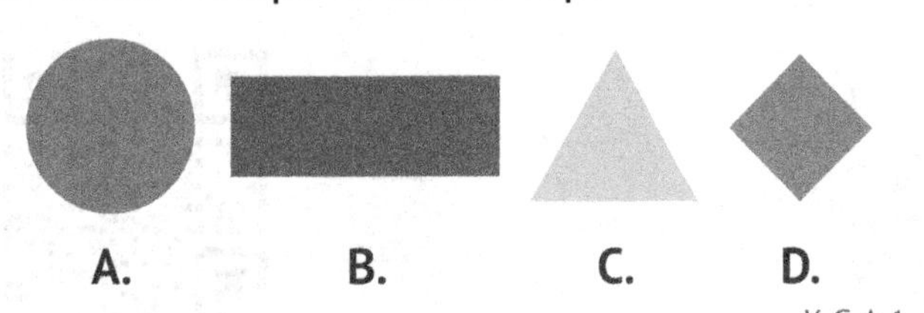

A. **B.** **C.** **D.**

K.G.A.1

3. Which shape is at the bottom?

A. **B.** **C.** **D.**

K.G.A.

4. Which shape is to the left?

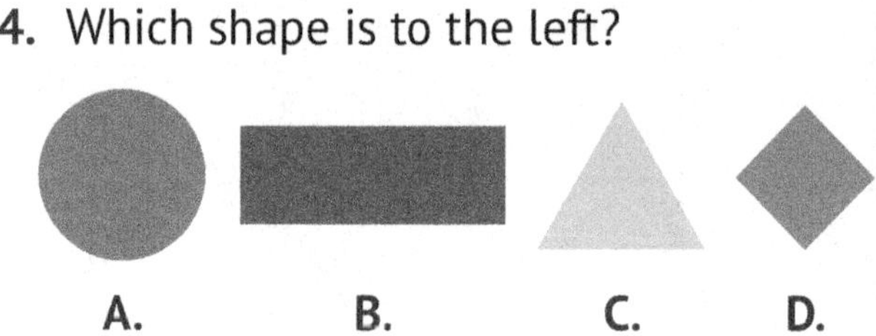

A. **B.** **C.** **D.**

K.G.A.

5. Which shape is to the right?

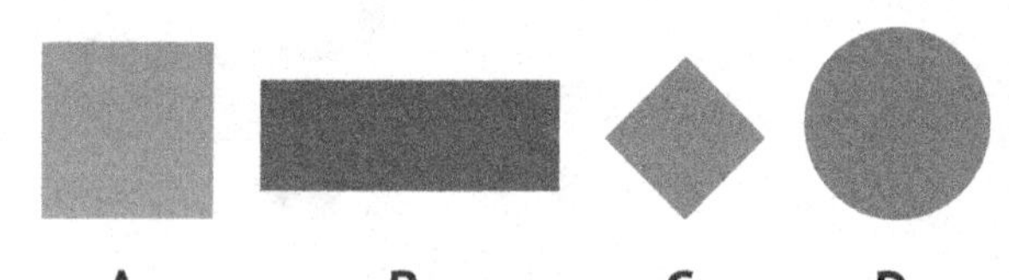

A. **B.** **C.** **D.**

K.G.A.

6. Which shape has the most?

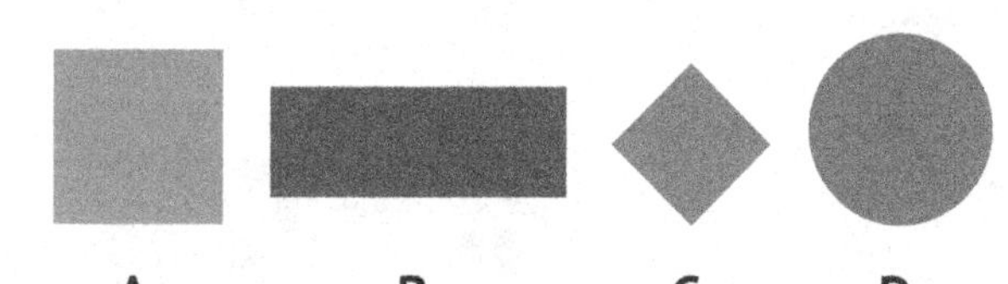

A. **B.** **C.** **D.**

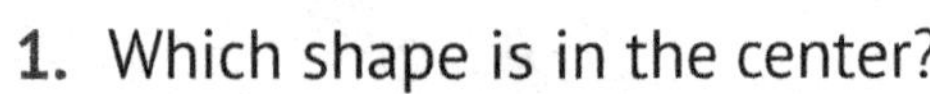

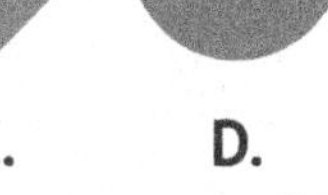

K.G.A.

TIP of the DAY

A circle does not have any corners or sides.

WEEK 18 : DAY 2

se the diagram below to answer the llowing questions.

. What is the name of the shape on the far left?

____

K.G.A.1

. What is the name of the shape on the far right?

____

K.G.A.1

3. What is the name of the shape that is pictured three times?

____

K.G.A.1

4. What is the name of the shape that is above the triangle?

____

K.G.A.1

5. How many triangles are there?

____

K.G.A.1

6. How many shapes are there in all?

____

K.G.A.1

A triangle has three corners and sides.

WEEK 18 : DAY 3

Use the diagram below to answer the following questions.

1. How many circles are there?

A. 1
B. 2
C. 3
D. 4

K.G.A.1

2. Which shape is above the triangle?

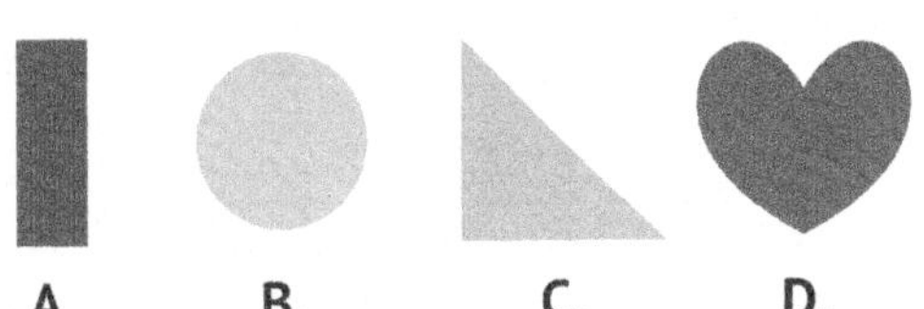

A. B. C. D.

K.G.A.1

3. Which shape is to the left of the circle?

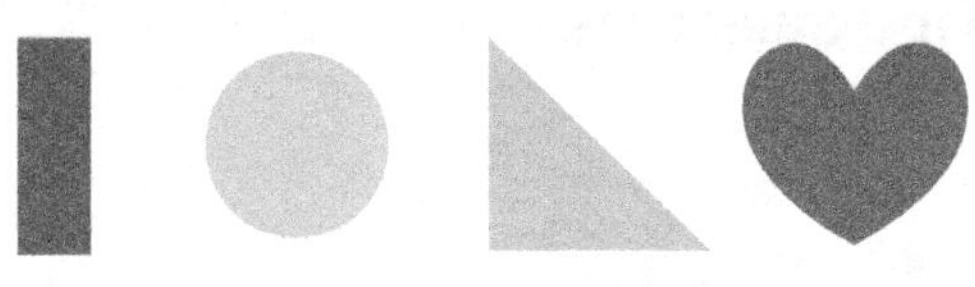

A. B. C. D.

K.G.A.

4. Which image is to the right of th triangle?

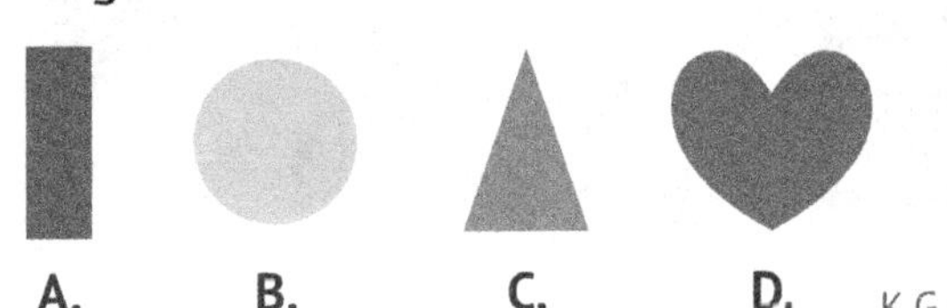

A. B. C. D. K.G.A.

5. Which image has the most?

A. B. C. D.

K.G.A.

6. How many shapes are there?

A. 5
B. 6
C. 7
D. 8

K.G.A.

A rectangle has 4 corners and sides.

WEEK 18 : DAY 4

se the diagram below to answer the
ollowing questions.

. Which shape is above the rectangle?

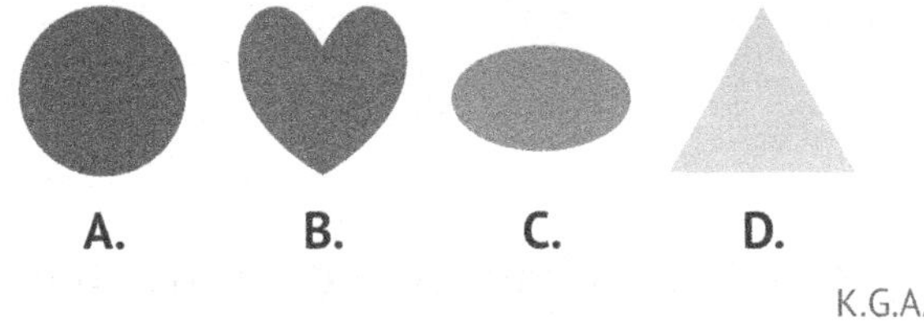

A. B. C. D.

K.G.A.1

. Which shape is below the triangle?

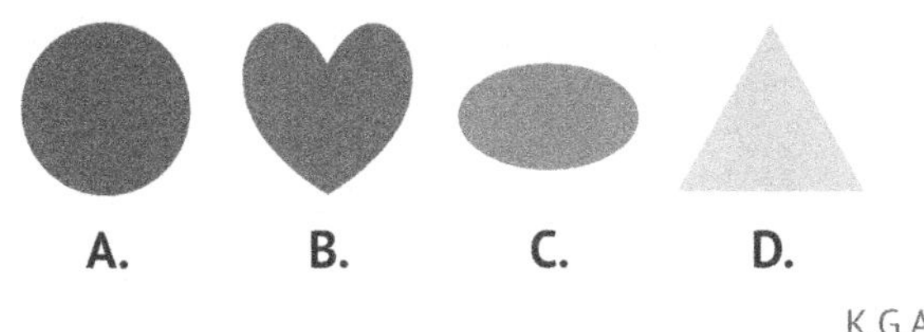

A. B. C. D.

K.G.A.1

3. Which shape is above the triangle?

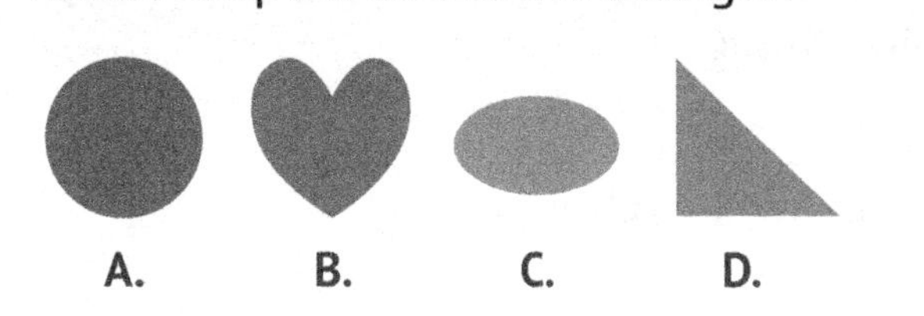

A. B. C. D.

K.G.A.1

4. Which shape is to the left of the oval?

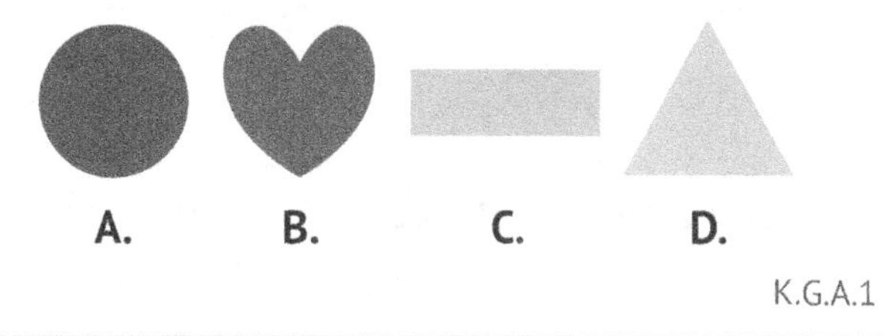

A. B. C. D.

K.G.A.1

5. Which shape is to the right of the two circles?

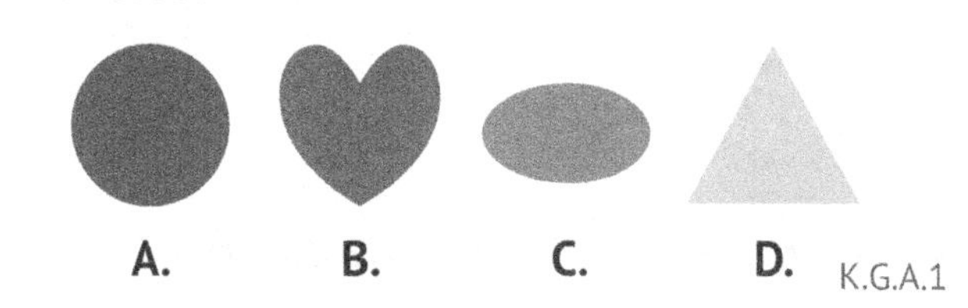

A. B. C. D. K.G.A.1

6. Which shape has three in the picture?

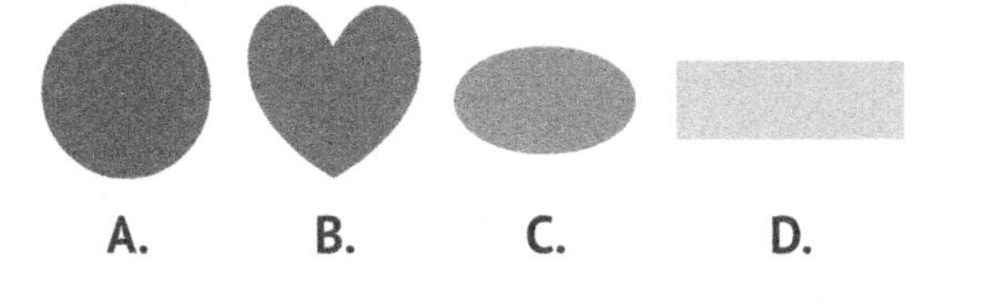

A. B. C. D.

K.G.A.1

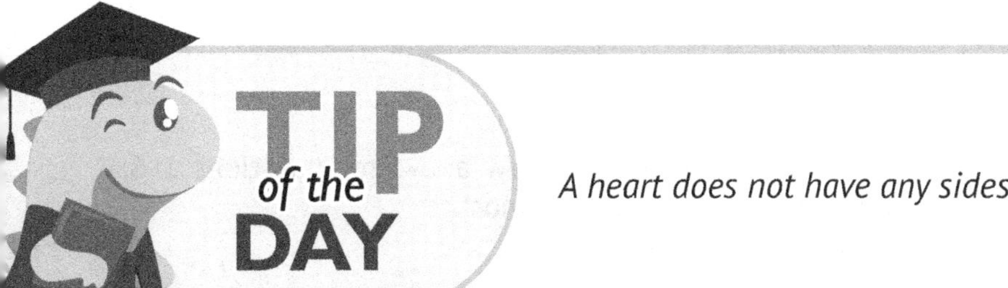

A heart does not have any sides.

WEEK 18 : DAY 5

ASSESSMENT

Create a picture based on the following directions.

1. Draw a circle.

K.G.A.1

2. Draw 2 hearts above the circle.

K.G.A.1

3. Draw 1 square under the circle.

K.G.A.1

4. Draw 3 triangles under the square.

K.G.A.

5. Draw 1 circle to the left of the hearts.

K.G.A.1

6. Draw 1 rectangle around all the shapes.

K.G.A.1

DAY 6
Challenge question

From all the shapes you drew above for questions 1-6, which shape is pictured the most?

1.MD.C.4

WEEK 19

ARGOPREP.COM

VIDEO
EXPLANATIONS

In week 19 you will need to correctly name shapes regardless of their orientations or overall size.

You can find detailed video explanations of each problem in the book by visiting:
ArgoPrep.com/ccmk

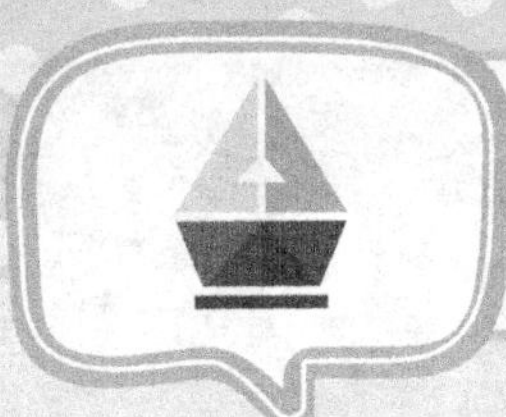

WEEK 19 : DAY 1

1. Circle the name of each pictured shape.

A. Circle
B. Square
C. Rectangle
D. Triangle

K.G.A.2

2. Circle the name of each pictured shape.

A. Circle
B. Square
C. Rectangle
D. Triangle

K.G.A.2

3. Circle the name of each pictured shape.

A. Circle
B. Square
C. Rectangle
D. Triangle

K.G.A.2

4. Circle the name of each pictured shape.

A. Circle
B. Square
C. Rectangle
D. Triangle

K.G.A.

5. Circle the name of each pictured shape.

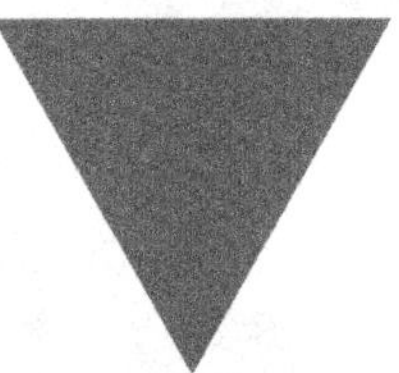

A. Circle
B. Square
C. Rectangle
D. Triangle

K.G.A.

6. Circle the name of each pictured shape.

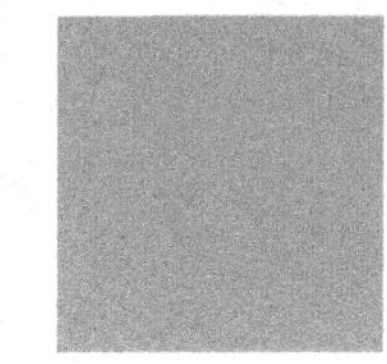

A. Circle
B. Square
C. Rectangle
D. Triangle

K.G.A.

A shape is a shape no matter the color or size.

WEEK 19 : DAY 2

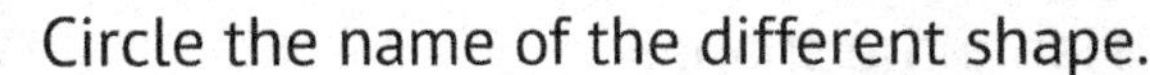

Circle the name of the different shape.

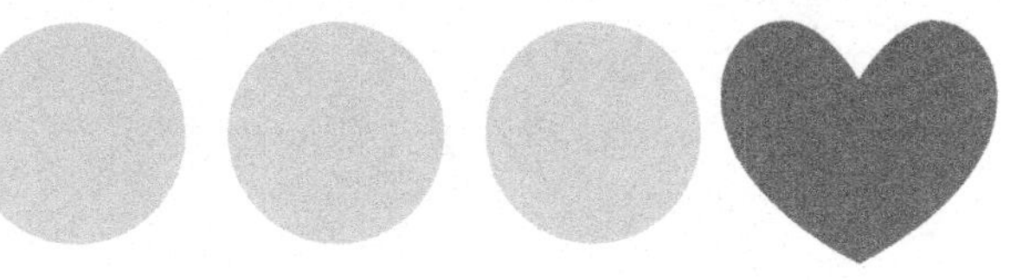

A. Heart
B. Triangle
C. Circle
D. Rectangle

K.G.A.2

Circle the name of the different shape.

A. Circle
B. Square
C. Rectangle
D. Triangle

K.G.A.2

Circle the name of the different shape.

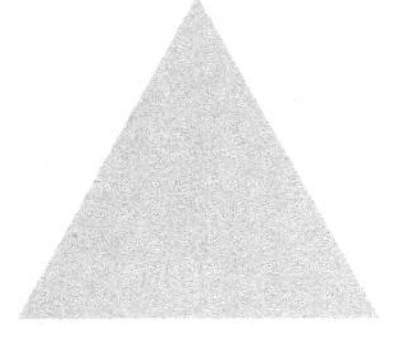

A. Circle
B. Square
C. Rectangle
D. Triangle

K.G.A.2

4. Circle the name of the different shape.

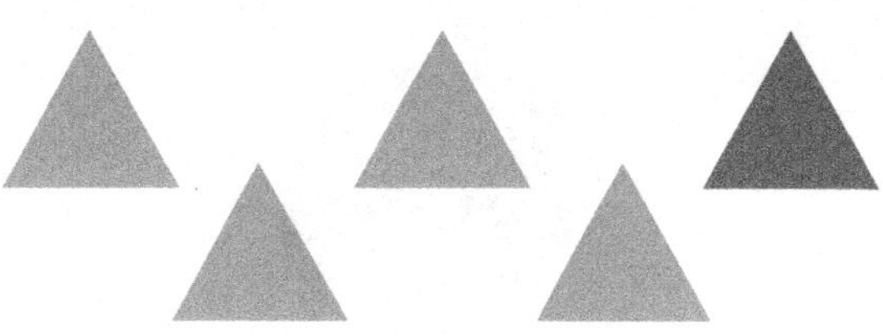

A. Circle
B. Square
C. Rectangle
D. Triangle

K.G.A.2

5. Circle the name of the different shape.

A. Circle
B. Square
C. Rectangle
D. Triangle

K.G.A.2

6. Circle the name of the different shape.

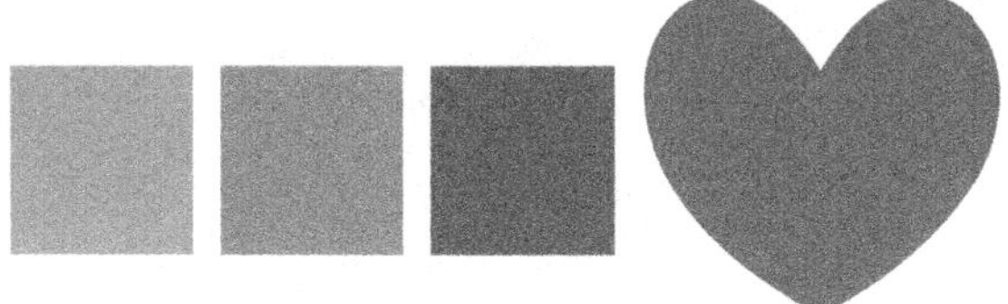

A. Circle
B. Square
C. Rectangle
D. Heart

K.G.A.2

Shapes are different based on their number of sides and corners.

WEEK 19 : DAY 3

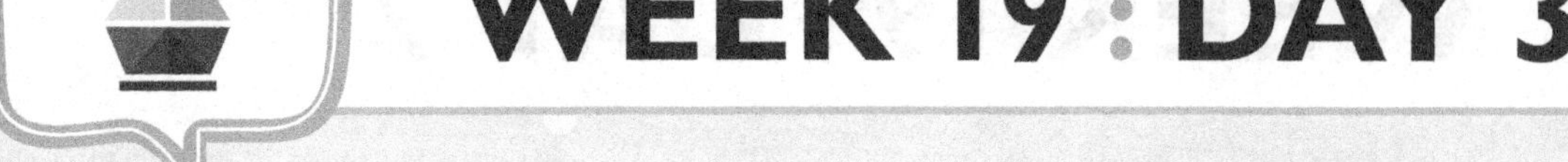

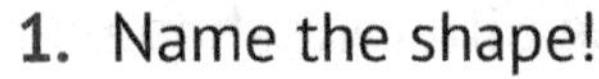

1. Name the shape!

_______________ K.G.A.2

2. Name the shape!

_______________ K.G.A.2

3. Name the shape!

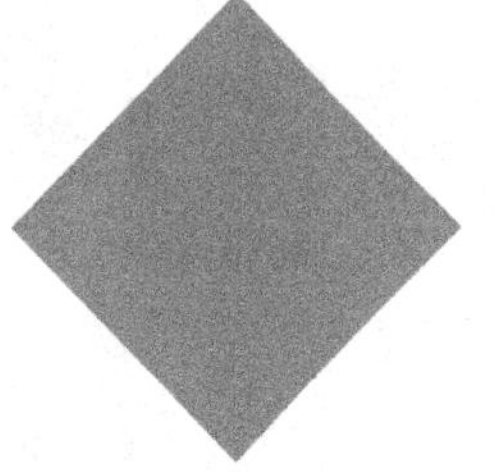

_______________ K.G.A.2

4. Name the shape!

_______________ K.G.A.

5. Name the shape!

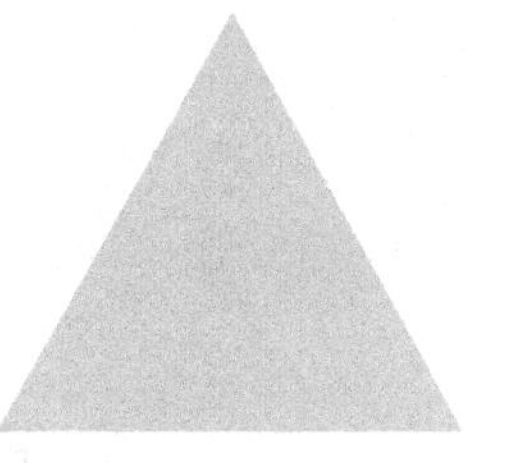

_______________ K.G.A.2

6. Name the shape!

_______________ K.G.A.2

Names of shapes: circle, triangle, heart, diamond, square, rectangle, oval

WEEK 19 : DAY 4

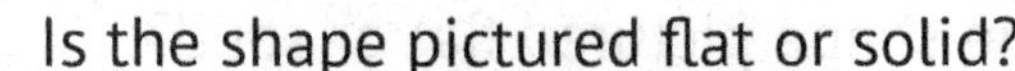

Is the shape pictured flat or solid?

______________ K.G.A.3

Is the shape pictured flat or solid?

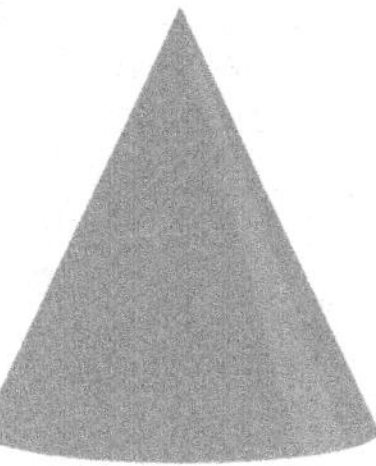

______________ K.G.A.3

Is the shape pictured flat or solid?

______________ K.G.A.3

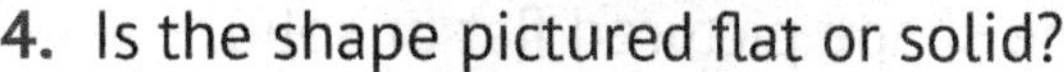

4. Is the shape pictured flat or solid?

______________ K.G.A.3

5. Is the shape pictured flat or solid?

______________ K.G.A.3

6. Is the shape pictured flat or solid?

______________ K.G.A.3

Know the differences between a circle, triangle, square, rectangle and pentagon.

WEEK 19 : DAY 5

ASSESSMENT

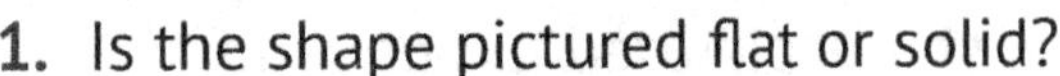

1. Is the shape pictured flat or solid?

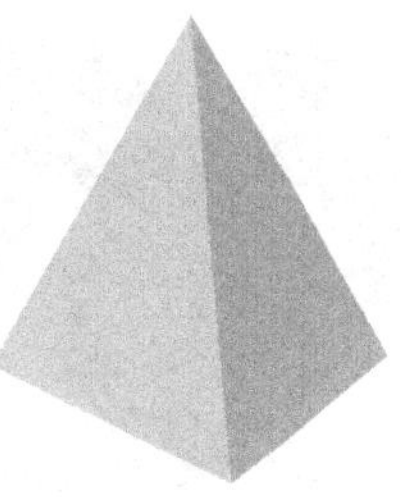

______________________ K.G.A.3

2. Is the shape pictured flat or solid?

______________________ K.G.A.3

3. Is the shape pictured flat or solid?

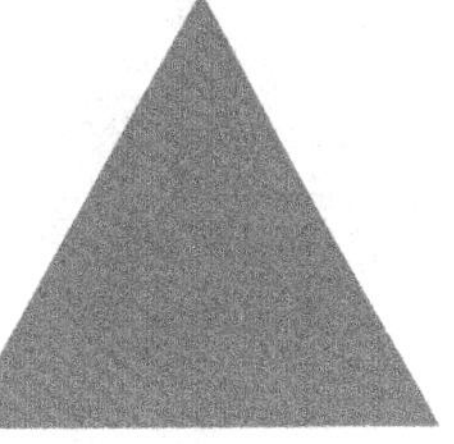

______________________ K.G.A.3

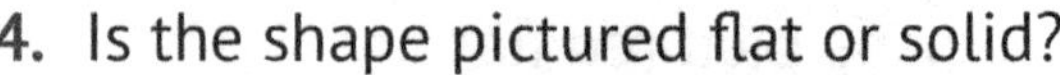

4. Is the shape pictured flat or solid?

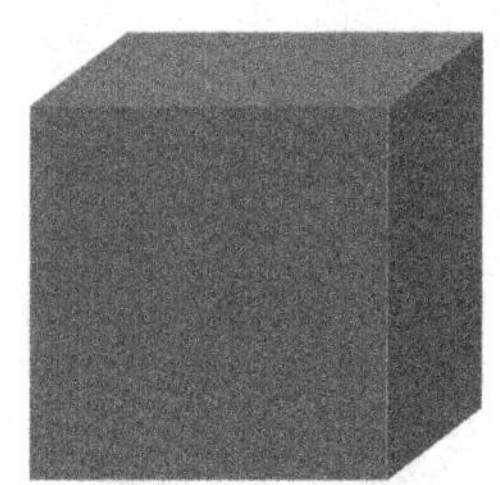

______________________ K.G.A.

5. Is the shape pictured flat or solid?

______________________ K.G.A.

6. Is the shape pictured flat or solid?

______________________ K.G.A.

DAY 6
Challenge question

Is a piece of paper flat or solid?

1.GA.1

Woah! Congratulations on making it this far. In week 20 you will analyze and compare two- and three-dimensional shapes.

You can find detailed video explanations of each problem in the book by visiting: ArgoPrep.com/ccmk

WEEK 20 : DAY I

1. Which shape has three sides?

   A. Square
   B. Triangle
   C. Circle
   D. Oval

   K.G.B.4

2. Which shape has no corners?

   A. Rectangle
   B. Circle
   C. Square
   D. Triangle

   K.G.B.4

3. Which shape has four sides?

   A. Circle
   B. Triangle
   C. Heart
   D. Square

   K.G.B.4

4. Which shape has all sides of the sam length?

   A. Square
   B. Rectangle
   C. Circle
   D. Heart

   K.G.B.

5. Which shape has two sides of the sam length?

   A. Circle
   B. Oval
   C. Triangle
   D. Rectangle

   K.G.B.

6. Which shape has four corners?

   A. Oval
   B. Circle
   C. Rectangle
   D. Triangle

   K.G.B.

Eliminate the shapes that do not have the stated characteristics first.

WEEK 20 : DAY 2

. Draw a shape that has 1 corner.

K.G.B.4

4. Draw a shape that has at least two sides the same length.

K.G.B.4

. Draw a shape that has no sides.

K.G.B.4

5. Draw a shape that has three corners.

K.G.B.4

. Draw a shape that has four corners.

K.G.B.4

6. Draw a shape that has six sides.

K.G.B.4

Draw the shape by focusing on its sides and corners first.

WEEK 20 : DAY 3

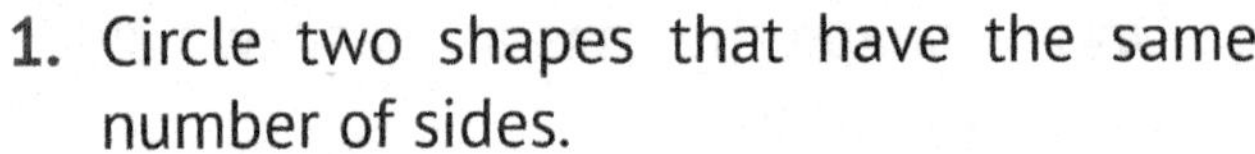

1. Circle two shapes that have the same number of sides.

K.G.B.4

2. Circle two shapes that have less than four corners.

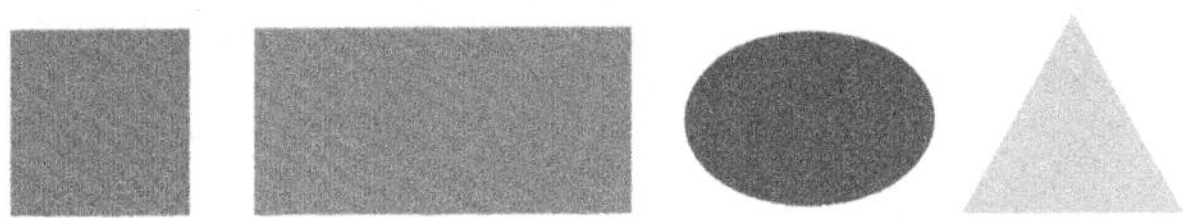

K.G.B.4

3. Circle two shapes that have four sides.

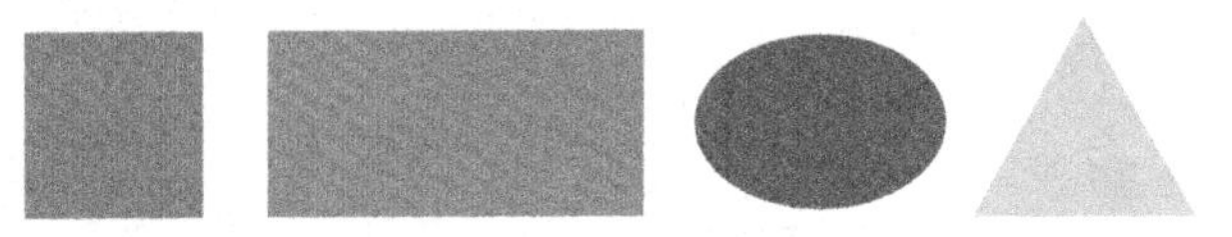

K.G.B.4

4. Circle two shapes that have at least thre corners.

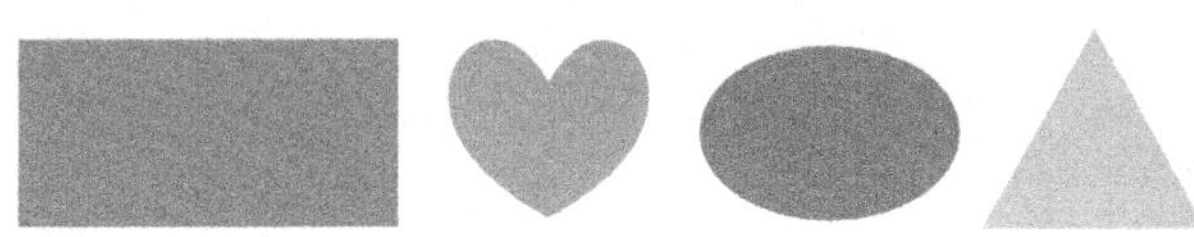

K.G.B.

5. Circle two shapes that have no corners.

K.G.B.

6. Circle two shapes that have at least thre sides.

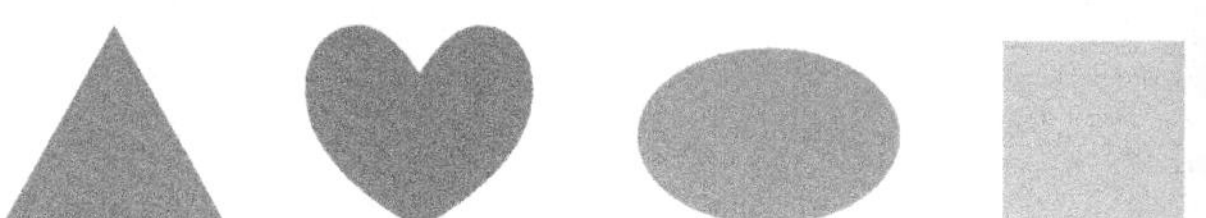

K.G.B.

Count the number of sides of each shape.

WEEK 20 : DAY 4

. Draw a picture of a circle.

K.G.B.5

4. Draw a picture of a square.

K.G.B.5

. Draw a picture of a triangle.

K.G.B.5

5. Draw a picture of an oval.

K.G.B.5

. Draw a picture of a rectangle.

K.G.B.5

6. Draw a picture of a heart.

K.G.B.5

Think of real life objects that fit the shape described.

WEEK 20 : DAY 5

ASSESSMENT

1. Draw a picture of two triangles that make a rectangle.

K.G.B.6

4. Draw a picture of four triangles that mak a diamond.

K.G.B.

2. Draw a picture of two squares that make a rectangle.

K.G.B.6

5. Draw a picture of two rectangles tha make a square.

K.G.B.6

3. Draw a picture of a square and a triangle that makes a house.

K.G.B.6

6. Draw a picture of an oval using a rectangl and two parts of a circle.

K.G.B.6

DAY 6
Challenge question

What is a real life example of an oval?

1.GA.2 & 1.GA.3

THE
END

Great job finishing all 20 weeks!
You should be ready for any test.

ASSESSMENT

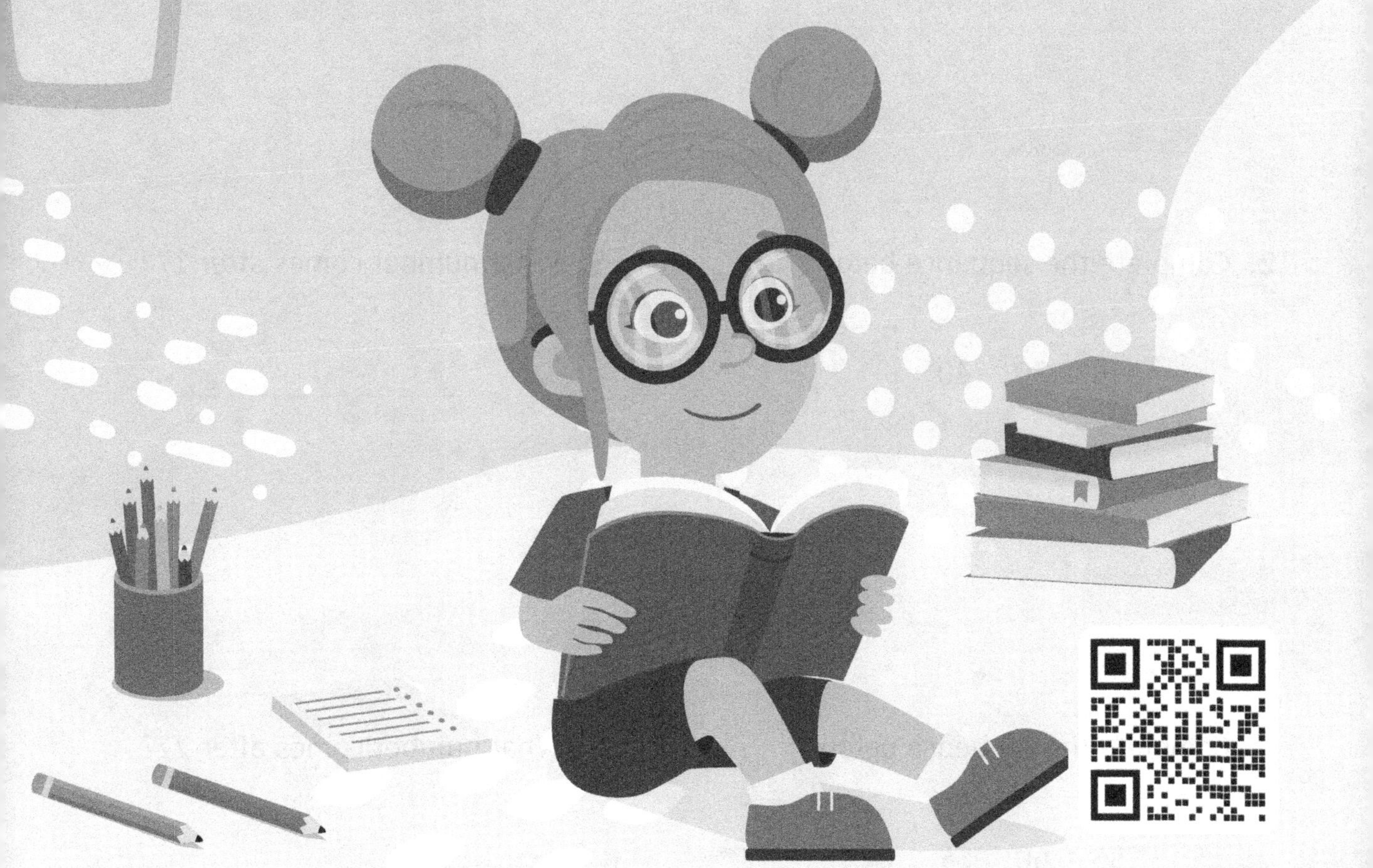

Try this assessment to see how much you've learned - good luck!

ASSESSMENT

1. Complete the sequence below:

14 15 16 ☐

2. Complete the sequence below:

20 30 40 ☐

3. Complete the sequence below:

55 60 65 ☐

4. What number comes after 6?

____

5. What number comes after 17?

____

6. What number comes after 22?

____

ASSESSMENT

Count the objects and write the number that would represent how many items there are.

____

Count the objects and write the number that would represent how many items there are.

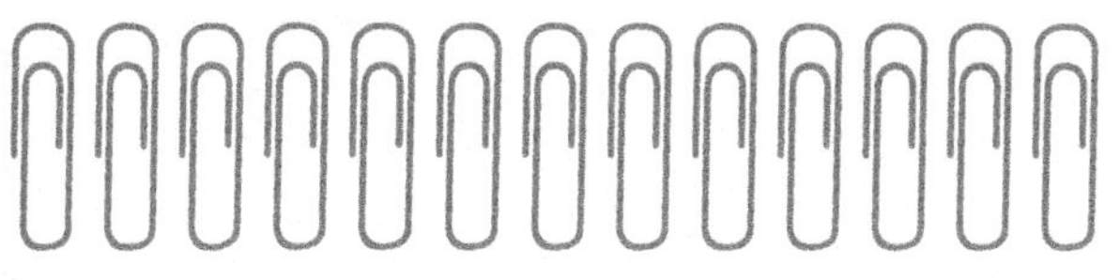

____

Write the numeral represented by the word "five"

____

10. Write the numeral represented by the word "eleven"

____

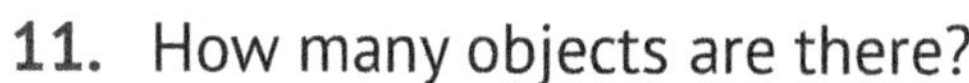

11. How many objects are there?

____

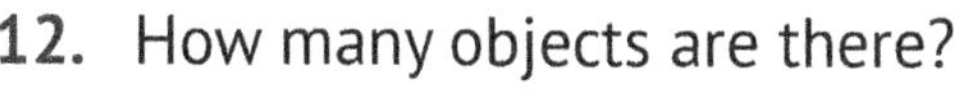

12. How many objects are there?

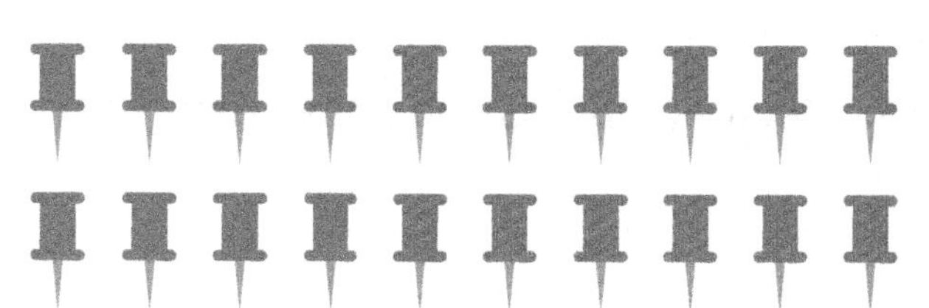

____

ASSESSMENT

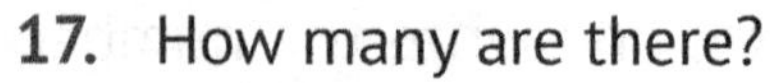

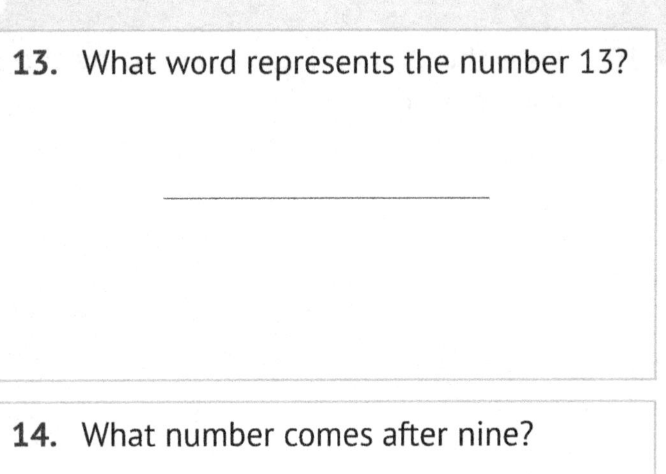

13. What word represents the number 13?

____

14. What number comes after nine?

____

15. What number is one larger than 3?

____

16. What number is one larger than 18?

____

17. How many are there?

____

18. Draw 4 number of objects.

19. Draw 9 number of objects.

ASSESSMENT

0. Which set is smaller?

____

23. Which number is larger? 5 or 2

____

1. Which set is larger?

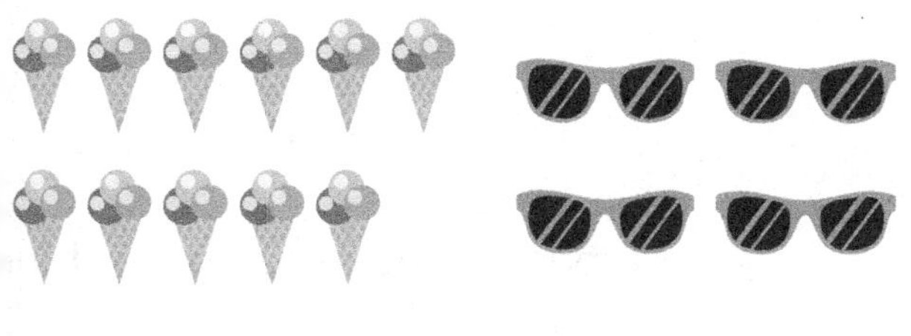

____

24. Which number is larger? 16 or 14

____

2. Which set is smaller?

____

25. Represent 4 + 3 with a drawing of 2 sets of objects.

ASSESSMENT

26. Draw 14 objects, cross out 12. How many objects are left?

27. Draw 4 objects. How many more objects should be drawn to reach 10?

28. Draw 9 objects, cross out 2. How many objects are left.

29. What number do you need to combin with 6 to make 10?

____

30. What number do you need to combin with 2 to make 8?

____

31. What number do you need to combin with 13 to make 17?

____

32. 5 + 3 = ☐

ASSESSMENT

3. 2 + 4 = ☐

4. 9 + 2 = ☐?

5. 7 + 1 = ☐

6. 6 + 3 = ☐

37. 2 + 5 = ☐

38. 1 + 4 = ☐

39. Draw 14 objects and cross out 8. How many objects are left?

ASSESSMENT

40. Draw 9 objects and cross out 4. How many objects are left?

41. What would you use to measure how tall a giraffe is?

____

42. What would you use to measure how heavy a backpack is?

____

43. What is heavier?

A. A bowling ball

B. A bouncy ball

44. What is shorter?

A. A pencil

B. A baby's shoe

45. Draw a set of shapes: 3 circles, 2 rectangles, 4 squares and a triangle.

ASSESSMENT

6. What shape is the most?

____

7. What shape is the least?

____

48. How many shapes have 4 corners?

____

49. How many shapes have 3 sides?

____

50. Draw a circle above a square.

51. Draw 2 rectangles that make a square.

ANSWER KEY

VIDEO
EXPLANATIONS

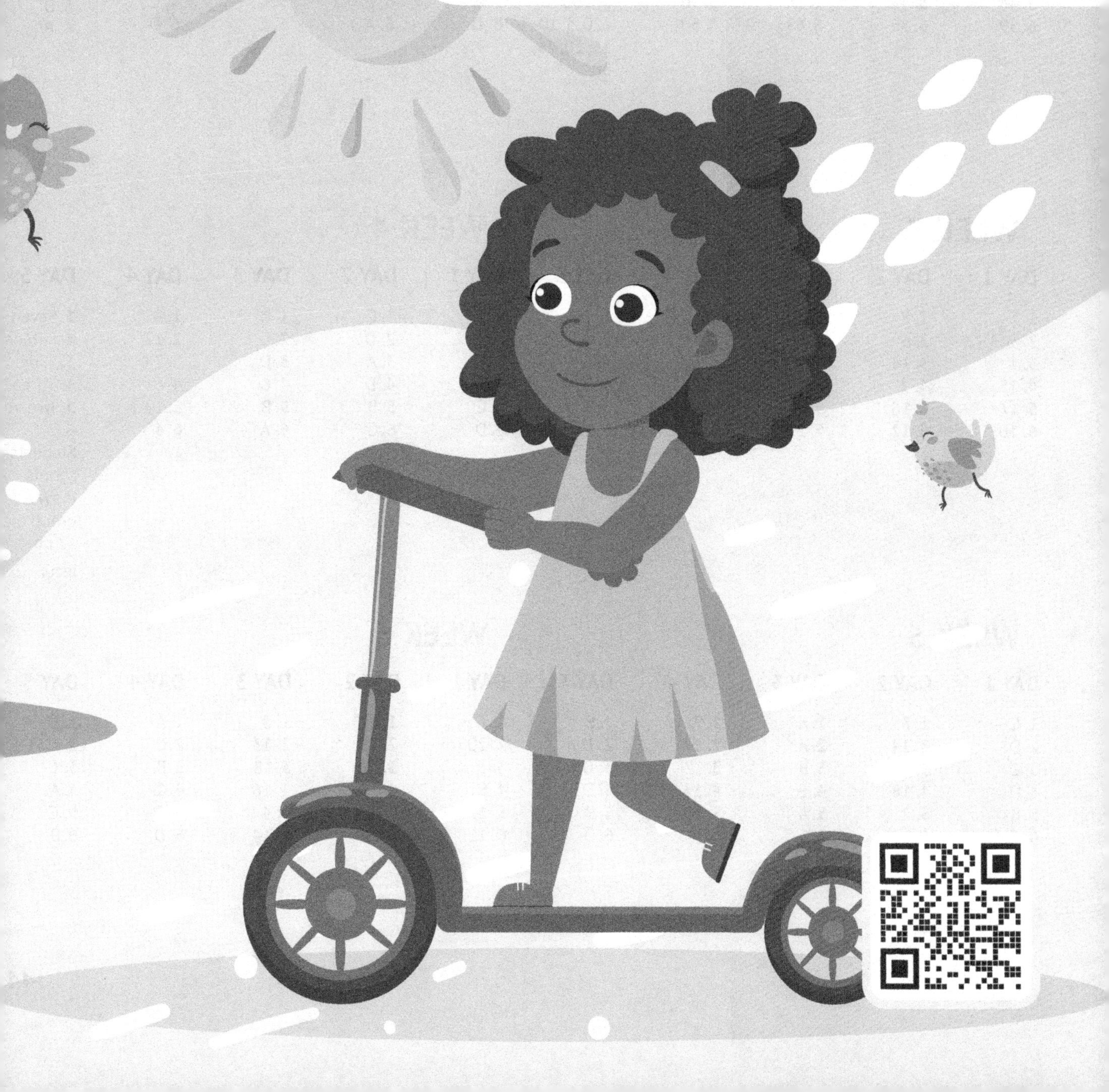

ANSWER KEY

WEEK 1

DAY 1	DAY 2	DAY 3	DAY 4	DAY 5
1. 5	1. 65	1. 42	1. 40	1. 17
2. 18	2. 82	2. 90	2. 60	2. 80
3. 31	3. 83	3. 53	3. 100	3. 30
4. 36	4. 97	4. 7	4. 70	4. 3
5. 44	5. 48	5. 14	5. 90	5. 100
6. 59	6. 26	6. 69	6. 50	6. 0, 5, 10, 15, 20, 25, 30

WEEK 2

DAY 1	DAY 2	DAY 3	DAY 4	DAY 5
1. A	1. C	1. A	1. C	1. C
2. C	2. A	2. C	2. C	2. A
3. D	3. B	3. D	3. B	3. B
4. B	4. C	4. C	4. A	4. A
5. A	5. B	5. C	5. C	5. D
6. D	6. A	6. C	6. D	6. B

WEEK 3

DAY 1	DAY 2	DAY 3	DAY 4	DAY 5
1. 7	1. 8	1. 9	1. 3	1. 10
2. 3	2. 6	2. 5	2. 7	2. 4
3. 4	3. 14	3. 13	3. 5	3. 8
4. 15	4. 2	4. 1	4. 1	4. 2
5. 17	5. 16	5. 18	5. 9	5. 6
6. 10	6. 12	6. 11	6. 11	6. 27

WEEK 4

DAY 1	DAY 2	DAY 3	DAY 4	DAY 5
1. B	1. C	1. B	1. 8	1. Seven
2. A	2. D	2. C	2. 12	2. Fifteen
3. C	3. A	3. D	3. 2	3. Nine
4. A	4. D	4. C	4. 6	4. Three
5. C	5. B	5. B	5. 10	5. Twelve
6. D	6. C	6. A	6. 4	6. Students should draw a group of 17 objects.

WEEK 5

DAY 1	DAY 2	DAY 3	DAY 4	DAY 5
1. C	1. 7	1. A	1. 9	1. B
2. D	2. 14	2. A	2. 17	2. D
3. C	3. 11	3. B	3. 12	3. D
4. D	4. 18	4. B	4. 13	4. D
5. B	5. 2	5. A	5. 10	5. B
6. C	6. 12	6. C	6. 3	6. D

WEEK 6

DAY 1	DAY 2	DAY 3	DAY 4	DAY 5
1. 7	1. 12	1. 2	1. A	1. B
2. 20	2. 16	2. 13	2. C	2. C
3. 3	3. 2	3. 18	3. B	3. C
4. 6	4. 19	4. 16	4. C	4. A
5. 17	5. 8	5. 4	5. C	5. C
6. 11	6. 11	6. 14	6. D	6. B

ANSWER KEY

WEEK 7

DAY 1	DAY 2	DAY 3	DAY 4	DAY 5
1. 3	1. 14	1. 1	1. Draw 4 objects.	1. Draw 7 objects.
2. 7	2. 20	2. 19	2. Draw 10 objects.	2. Draw 3 objects.
3. 13	3. 4	3. 11	3. Draw 6 objects.	3. Draw 12 objects.
4. 2	4. 17	4. 2	4. Draw 2 objects.	4. Draw 9 objects.
5. 1	5. 6	5. 15	5. Draw 11 objects.	5. Draw 5 objects.
6. 10	6. 12	6. 8	6. Draw 8 objects.	6. Draw 13 objects.

WEEK 8

DAY 1	DAY 2	DAY 3	DAY 4	DAY 5
1. B	1. B	1. Circle 2 objects.	1. B	1. B
2. A	2. A	2. Circle 13 objects.	2. A	2. A
3. B	3. B	3. Circle 7 objects.	3. B	3. A
4. A	4. B	4. Circle 5 objects.	4. B	4. B
5. B	5. A	5. Circle 16 objects.	5. A	5. B
6. B	6. A	6. Circle 10 objects.	6. A	6. B

WEEK 9

DAY 1	DAY 2	DAY 3	DAY 4	DAY 5
1. 3	1. 2	1. 9	1. 4	1. 2
2. 9	2. 4	2. 9	2. 3	2. 10
3. 9	3. 6	3. 7	3. 1	3. 5
4. 8	4. 3	4. 8	4. 4	4. 1
5. 6	5. 7	5. 9	5. 6	5. 5
6. 5	6. 5	6. 10	6. 3	6. 2

WEEK 10

DAY 1	DAY 2	DAY 3	DAY 4	DAY 5
1. A	1. A	1. 1	1. 2	1. 5
2. B	2. C	2. 4	2. 2	2. 7
3. B	3. A	3. 2	3. 1	3. 3
4. B	4. B	4. 6	4. 4	4. 7
5. D	5. D	5. 4	5. 3	5. 14
6. B	6. C	6. 6	6. 1	6. 4

WEEK 11

DAY 1	DAY 2	DAY 3	DAY 4	DAY 5
1. Draw a model.	1. Draw a model.	1. Draw a model.	1. Draw a model.	1. Draw a model.
2. 4+2	2. 8-3	2. 3+4	2. 3+6	2. 6+4
3. 6	3. 5	3. 7	3. 9	3. 10
4. Draw a model.	4. Draw a model.	4. Draw a model.	4. Draw a model.	4. Answers will vary.
5. 6-3	5. 10-6	5. 6+2	5. Sister	5. Answers will vary.
6. 3	6. 4	6. 8	6. 4	6. Answers will vary.

WEEK 12

DAY 1	DAY 2	DAY 3	DAY 4	DAY 5
1. 3	1. 5	1. 3	1. B	1. A
2. 1	2. 2	2. 5	2. B	2. D
3. 1	3. 3	3. 4	3. B	3. B
4. 1	4. 2	4. 6	4. A	4. A
5. 1	5. 5	5. 1	5. D	5. Draw 4 objects and circle 2.
6. 6	6. 2	6. 1	6. C	6. C

ANSWER KEY

WEEK 13

DAY 1	DAY 2	DAY 3	DAY 4	DAY 5
1. 2	1. 1	1. 3	1. C	1. B
2. 8	2. 8	2. 1	2. C	2. D
3. 7	3. 4	3. 7	3. C	3. A
4. 5	4. 5	4. 4	4. C	4. B
5. 6	5. 2	5. 9	5. A	5. Draw 10 objects and circle 2 groups of 5.
6. 9	6. 3	6. 6	6. C	6. 10

WEEK 14

DAY 1	DAY 2	DAY 3	DAY 4	DAY 5
1. 3	1. 8	1. 6	1. B	1. C
2. 7	2. 9	2. 12	2. A	2. B
3. 3	3. 6	3. 7	3. C	3. A
4. 7	4. 11	4. 9	4. D	4. C
5. 8	5. 5	5. 6	5. B	5. C
6. 6	6. 4	6. 6	6. C	6. D

WEEK 15

DAY 1	DAY 2	DAY 3	DAY 4	DAY 5
1. 7	1. 6	1. 17	1. 2	1. 4
2. 14	2. 11	2. 8	2. 2	2. 4
3. 13	3. 5	3. 7	3. 3	3. 2
4. 4	4. 8	4. 8	4. 1	4. 3
5. 8	5. 2	5. 6	5. 2	5. 5
6. 12	6. 7	6. 9	6. 3	6. 18

WEEK 16

DAY 1	DAY 2	DAY 3	DAY 4	DAY 5
1. Length	1. Length	1. Bookshelf	1. Elephant	1. Draw
2. Weight	2. Weight	2. Cake	2. Sofa	2. Draw
3. Weight	3. Length	3. Cat	3. Carrot	3. Draw
4. Length	4. Weight	4. Snowman	4. Laptop	4. Draw
5. Length	5. Length	5. Mouse	5. Pants	5. Draw
6. Weight	6. Length	6. Car	6. Cow	6. Draw

Challenge Question

Week 1: 25.
Week 2: 47.
Week 3: eighty-two.
Week 4: 106.
Week 5: 38.
Week 6: 72.
Week 7: 49
Week 8: Students should draw a set with at least 10 objects.
Week 9: Any number greater than 26.
Week 10: 4.
Week 11: No.
Week 12: 8.
Week 13: 6.
Week 14: 8.
Week 15: Answers may vary but should include 3 numbers.
Week 16: Street.
Week 17: 12:30.
Week 18: Triangle.
Week 19: Flat.
Week 20: Answers may vary.

ANSWER KEY

WEEK 17

DAY 1	DAY 2	DAY 3	DAY 4	DAY 5
1. C	1. B	1. C	1. A	1. Add 3 suns to the set.
2. C	2. D	2. B	2. A and D	2. C
3. C	3. A abd C	3. D	3. C	3. A
4. B	4. B and D	4. B	4. B	4. C
5. A and D	5. A	5. B	5. A	5. C
6. B	6. D	6. C	6. A and D	6. A

WEEK 18

DAY 1	DAY 2	DAY 3	DAY 4	DAY 5
1. B	1. Rectangle	1. C	1. A	1. Draw
2. C	2. Heart	2. C	2. B	2. Draw
3. D	3. Circle	3. A	3. C	3. Draw
4. A	4. Triangle	4. D	4. A or C	4. Draw
5. A	5. 2	5. B	5. C	5. Draw
6. A	6. 7	6. C	6. D	6. Draw

WEEK 19

DAY 1	DAY 2	DAY 3	DAY 4	DAY 5
1. A	1. A	1. Circle	1. Flat	1. Solid
2. B	2. C	2. Square	2. Solid	2. Flat
3. D	3. D	3. Diamond	3. Solid	3. Flat
4. C	4. D	4. Rectangle	4. Flat	4. Solid
5. D	5. A	5. Triangle	5. Solid	5. Solid
6. B	6. D	6. Heart	6. Flat	6. Solid

WEEK 20

DAY 1	DAY 2	DAY 3	DAY 4	DAY 5
1. B	1. Draw	1. A and D	1. Draw	1. Draw
2. B	2. Draw	2. C and D	2. Draw	2. Draw
3. D	3. Draw	3. A and B	3. Draw	3. Draw
4. A	4. Draw	4. A and D	4. Draw	4. Draw
5. D	5. Draw	5. B and C	5. Draw	5. Draw
6. C	6. Draw	6. A and D	6. Draw	6. Draw

Assessment

1. 17
2. 50
3. 70
4. 7
5. 18
6. 23
7. 6
8. 13
9. 5
10. 11
11. 8
12. 20
13. Thirteen
14. 10
15. 4
16. 19
17. 12
18. Students should draw 4 objects.
19. Students should draw 9 objects.
20. 3
21. 11
22. 1
23. 5
24. 16
25. Students should draw objects that represent the problem.
26. 2
27. 6
28. 7
29. 4
30. 6
31. 4
32. 8
33. 6
34. 11
35. 8
36. 9
37. 7
38. 5
39. 6
40. 5
41. Length
42. Weight scale
43. Bowling ball
44. Baby's shoe
45. Students should draw 3 circles, 2 rectangles, 4 squares and a triangle.
46. Square
47. Triangle
48. 6
49. 1
50. Students should draw a circle above a square.
51. Students should draw 2 rect-angles that make a square.

KIDS WINTER ACADEMY

Kids Winter Academy by ArgoPrep covers material learned in September through December so your child can reinforce the concepts they should have learned in class. We recommend using this particular series during the winter break. These workbooks include two weeks of activities for math, reading, science, and social studies. Best of all, you can access detailed video explanations to all the questions on our website.

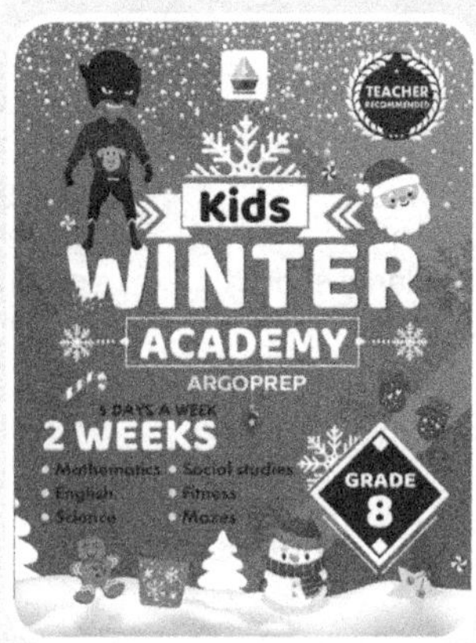

Kindergarten

DIPLOMA

The certificate is presented to:

____

your name

____

by school name

for successful completion of ArgoPrep's Kindergarten workbook

______________ ______________

Date *Signature*

Excellent work!

ARGOPREP

Made in the USA
Monee, IL
01 August 2024

62981186R00085